WALKING
—WITH—
VISWA

WALKING —WITH— VISWA

A
JOURNEY TO LEADERSHIP
ALONG THE FARMLANDS

G. BALASUBRAMANIAN

KONARK

Konark Publishers Pvt. Ltd
206, First Floor
Peacock Lane, Shahpur Jat
New Delhi - 110 049
+91-11-4105 5065
india@konarkpublishers.com, us@konarkpublishers.com
www.konarkpublishers.com

Copyright © G. Balasubramanian, 2025

All rights reserved. No part of this book may be reproduced or utilized in any form or by any means, electronic or mechanical, including photocopying, recording, or by any information storage and retrieval system, without prior written permission from the author or the publisher. The views and opinions expressed in this book are solely those of the author. While the accuracy of the facts, as reported by the author, has been verified to the fullest extent possible, the publisher is not liable in any way for the content.

ISBN: 978-81-987405-7-1

Edited by Dipali Singh
Jacket design by Sourish Mitra
Illustrations by Jitendra Kumar Patra
Typeset by Saanvi Graphics, Noida
Printed and bound in India by Anvi Composers

Dedicated
to
the Farmers of Bharat

Contents

	Acknowledgements	ix
	Praise	xi
	Prologue: Why did I Write this Book?	xv
	An Invite to an Experience	xix
1.	Meet My Friend Viswa	1
2.	Nurturing the Seed and the Self	10
3.	Mentoring the Farm and the Leader	21
4.	The Growing Seed and the Growing Leader	31
5.	Farming: A Lesson in Inclusive Leadership	45
6.	The Soil is the Universal Mother	59
7.	The Power of Seeds: Ownership Values	68
8.	Farmers: The Barefoot Leaders	79
9.	Leaders: Walk the Talk	91
10.	Farming and Leading in Crisis and Chaos	106
11.	Braving the Storms: The Farmer and the Leaders	120

12.	Problem-Solving on the Riverside	132
13.	Farmers and Leaders: Architects of Hope	144
14.	Resolving Conflicts in Physical and Human Nature	155
15.	The Glory of Agricultural Leadership	165
16.	Farmers and Leaders Celebrate Diversity	174
17.	Leadership: A Journey of Continuous Learning	185
18.	Celebrating the Leader in the Self	195
19.	Gratitude: The Way of Life	203
	Epilogue	215
	About the Author	218

Acknowledgements

I salute the benign providence that ushered the idea, the concept, the words and the energy into me to author this book. Without the Divine will, it would certainly not have happened.

I bow to Mother India, whose benevolence has blessed us with a gift of nature that is unparalleled in the world. As the beautiful Sanskrit saying goes, "*Janani Janmabhoomishcha Swargadapi Gariyasi*", meaning "Mother and Motherland are superior even to heaven". My travels through the length and breadth of the country gave me a deep understanding of her manifestation in all five elements – the earth, water, air, sky, and fire. From the towering Himalayas to the tip of Kanyakumari, her beauty is exceptional. The plurality of the Indian culture, embedded in her villages, has always been her strength.

I thank all my family members, who have always supported my endeavours. Their faith in me and my pursuits, and their patience in allowing me the space to follow my heart are commendable. Thank you all!

I am grateful to my friends, Giri Balasubramaniam (Pickbrain) and the young entrepreneur Ms Aisshwarya D.K.S. Hegde, who have always encouraged me to scale more heights.

Mr K.P.R. Nair, Managing Director of Konark Publishers Pvt. Ltd, has been a long-standing friend and one of the country's leading publishers. His willingness to publish this book eased many of my worries. His belief in my work and his guidance throughout the process were truly invaluable. I thank him profusely for his support.

I am deeply grateful to Ms Dipali Singh for her thoughtful editing of the manuscript. My heartfelt thanks also go to Coordinating Editor Jiza Joy, whose intellectual, linguistic, and professional competence helped bring this book to life in its truest form. I thank the entire Konark team who gave their valuable time and attention to this book.

It is difficult to thank everyone who has been part of this long journey in words. But I just want to assure them that they are always in my heart, and I will continue to need their support in my future walks, especially along the farmlands that remain close to my soul.

Thank you all!

Praise

Walking with Viswa leads you through a vivid landscape, as if stepping into a magnificent painting. Like the blending of rainbow colours, we experience the melding of life, as each page draws freely from ancient oriental scriptures, universal truths and recent breakthroughs in physics, technology, psychology, neuroscience, life sciences and life skills. These diverse elements are woven together to create brilliant parallels between profound management concepts and the simple, rustic wisdom and practices of agrarian life.

Just when we begin to relax and immerse ourselves in a story, we suddenly pause—a lofty saying drifts past. We go back to ruminate, to chew on its wisdom and to linger in its thought before turning the page and moving on. The author's deftness in packing so much wisdom so effortlessly is truly amazing.

Walking with Viswa is a book to savour and revisit—a source of insight to gain and regain in manifold ways.

Dr VASANTHI VASUDEV
Educator, author and poet

Walking with Viswa is a thoughtful exploration of leadership principles rooted in the context of village life. What sets this book apart is its unique approach: internalising and distilling leadership lessons through the lived experiences of rural communities. Written in an engaging, conversational style, the narrative draws readers in, making it both an enjoyable and thought-provoking read from start to finish.

Each chapter focuses on specific leadership principles, presented in a manner that is both accessible and relevant to a broad audience, particularly those interested in Indian leadership philosophy. The book has the potential to become a meaningful contribution to the field of leadership studies, especially within the framework of Indian culture.

The insights shared throughout serve not only as tools for personal growth and empowerment but also as powerful reminders of the values that underpin responsible and ethical citizenship.

A. AMARENDER REDDY
Joint Director, Policy Support Research, ICAR-NIBSM, Raipur

In this evocative narrative, the author returns to his cherished theme of leadership, framing it this time within the context of farming. This innovative approach not only underscores the inherent nobility of agriculture but also elevates the stature of those who toil in the fields.

The conversational tone, complemented by reflective questions at the end of each chapter, deepens reader engagement and encourages introspection.

India's heart beats in her villages, and this book offers a heartfelt tribute to these rural sanctuaries. It celebrates the farmers, the lessons drawn from farming and the multifaceted dimensions of leadership that resonate across all spheres of life.

Through this work, the author skilfully harvests a rich tapestry of insights, offering readers a profound understanding of the true essence of leadership.

A must-read for every leader and for anyone who aspires to become one.

P.C. BALASUBRAMANIAN
Entrepreneur, author and speaker

Walking with Viswa is a modern fable that offers timeless wisdom. It invites you to discover the many "Viswas" in your life while nurturing a solution-oriented mindset. This classic serves as a practical toolkit for success, drawing lessons from real conversations and experiences—the kind they don't teach at Harvard!

GIRI BALASUBRAMANIAM (PICKBRAIN)
Founder & CEO, Greycaps; Co-founder, Teacher Tribe

Prologue
Why Did I Write This Book?

Since my school days, I've grown tired of hearing my teachers, friends and other professionals say, "Look at Germany, Japan, and the US; look at all they've achieved and the kind of leadership they're providing to the world". I never challenged their teachings or perceptions. I simply accepted their statements without question. While my admiration for these countries grew immensely, and perhaps disproportionately, it also gave rise to a sense of inadequacy about myself, my people and my country.

Still, I held on to the dream that one day, someone would say, "Look at Bharat." Every morning, as I listened to All India Radio, a melodious tune would play, followed by the stirring song "*Vande Mataram! Sujalam, suphalam, malayaja shitalam, Shasyashyamalam, Mataram*", praising the laurels of Bharat's flora and fauna. This reinforced my hope that this dream would one day become a reality.

Later in my life, I had the opportunity to visit all these countries and many others. Reflecting on my personal experiences, I came to realise that many of the leadership principles celebrated abroad were already present in our own country, particularly in our villages. As an educator, I travelled across the length and breadth of the country and observed that local communities exhibited a deep sense of brotherhood and mutual respect in spite of the immense diversity they embraced. Local leadership displayed wisdom and courage in promoting cohabitation and ethical values, guiding people towards growth and peaceful coexistence. The common thread everywhere was a humanistic approach.

Throughout my professional career, I had the privilege of interacting with several senior faculty members from renowned management and leadership institutes. While most of them displayed varying dimensions of scholarship and an unparalleled understanding of critical issues related to management and leadership, I, as an outsider, observed all this with awe and admiration. I came to realise that many of these concepts were already in practice, albeit to varying degrees, in our villages.

This perspective was further reinforced during my post-retirement interactions with several members of the corporate world. Although the terminology, styles, approaches and practices differed—being more elementary, unstructured and culturally coherent—they were also strikingly authentic, empowering and empathetic. The framework that supported these practices in the villages of Bharat was rooted in social values and social consciousness.

It seemed crucial to convey the message that, despite her differences with other countries, India possesses an inherent and latent strength to navigate the heterogeneous and humane complexities native to her. Bharat, in its essence, has the ability to face its unique challenges. Therefore, any attempt to downplay the local model of leadership or deem it inferior to other global leadership styles or to those preached from the ivory towers of academia must be put to rest.

The saying "India lives in her villages" is more than just a statement; it is a reality. Time and again, the strength of Indian villages has been demonstrated to the world, not only through the remarkable progress of the Green Revolution but also through the adoption of democratic principles and governance at the grassroots level. Many of the country's luminaries in literature and other fields such as music, theatre, sports, art, science and technology spent their formative years in these villages, disproving the notion that wisdom is exclusive to iconic learning institutions.

Moreover, the fact that they emerged from a backdrop of diverse religions, castes, communities, languages and cultures stands as proof of the participative and pragmatic leadership that thrives in rural India. In recent times, the villages of Bharat have once again shown their resilience and adaptability—this time by embracing technology with ease, especially through initiatives like "Digital India".

I have had a few friends who, after achieving positions of importance in the corporate world, sought refuge in the villages, searching for peace, health and positivity. Yet, there is a noticeable trend of migration from villages, especially among the younger generation, who leave in pursuit of greener pastures, as the "greens" of agricultural lands and farms are not celebrated with the dignity they deserve. Perhaps their understanding of respect and dignity is somewhat tied to "modernity"—a belief system shaped by consumerist attitudes and sustained by commercial interests. However, this perception appears to be changing.

This book is an earnest attempt to restore faith in our villages and in early Indian practices, which are built on cultural camaraderie and compassionate cohabitation. It advocates recognising the leader within every individual and helping them realise "the power of the self". My intention is not to dismantle any existing theories, definitions, strategies or approaches to leadership proposed by individuals or institutions; rather, I seek to celebrate a perspective that has been integral to the Indian consciousness for centuries—honouring the *Annadhatha* (food-giver)—the farmer—and the lessons we can learn from them.

Some of my views may be open to debate, but as long as they serve as food for reflection and help us reconnect with the spirit of "Bharat" within us, I will be content.

I have consciously adopted a conversational model to allow for the "free flow of thoughts" through arguments, conversations, questions and engagements so that diverse perceptions of leaderships can be expressed and deliberated. In doing so, I hold close to my heart the proverb, "It takes a village to raise a leader".

Finally, each chapter includes statements that encourage readers to reflect on the essentials of leadership as well as engage with questions that may trigger some action. Enjoy reading!

Chennai   **G. Balasubramanian**
10 February 2025

An Invite to an Experience

Welcome to a wonderful journey! You and I are about to walk with Viswa, my friend, philosopher and guide. To give you a brief introduction, Viswa left the corporate world and settled in his village to follow his true passion: farming. In just a few years, he has transformed the village he now calls home. But honestly, that is just an insignificant part of his story. He breathes farming and sees life in every single farming operation. He views everything around him as a living organism, teaching him—and all of us—valuable lessons about life and leadership.

In this journey, we are accompanied by a team of young agricultural graduates—Amar, Shyam and Ankita—who are learning firsthand from Viswa what pragmatic leadership truly means. They carry a fire in their bellies, always spotting opportunities for engagement and entrepreneurship in those farmlands. Yet, they share as much love for literature as they do for technology and social work. From time to time, they will keep us, along with Viswa, engaged with their insightful questions and innovative responses.

We will also meet Muni, a middle-aged farmer who never went to high school. Viswa calls him his *guru* and "the leader of leaders". Muni's outlook on life is practical, purposeful and positive. Through his interventions, we will gain unique insights into what true leadership and purposeful living really mean.

I will be with you throughout the journey, learning at every step and seeking to understand the truth about leadership. You might ask why. Well, friends, leadership is not just what we read in textbooks. In reality, many of those concepts are interpretations of how people have managed their lives and relationships. Leadership is about understanding how they organise their daily affairs with ethics and insight. It is experiential learning, and through my own journey, I have discovered that cultural camaraderie plays a vital role in experiencing and evolving as a leader.

Finally, the farmers in the village, silent leaders who demonstrate leadership in many of their activities, from seeding to harvesting, will show us how to evolve as leaders without stress. They remind us that they practice leadership just as much as anyone in the corporate world. Unfortunately, they are not adequately celebrated!

This is a book with a difference. It doesn't preach. Instead, it simply brings you the breeze of leadership styles from the farmlands. Enjoy the journey!

Walk with Viswa to learn about life and leadership.

Thank you for your company.

G. Balasubramanian

*A farm includes the passion of the farmer's heart,
the interest of the farm's customers,
the biological activity in the soil,
the pleasantness of the air about the farm —
it's everything touching, emanating from,
and supplying that piece of landscape.
A farm is virtually a living organism.*

JOEL SALATIN
American farmer, lecturer and author

Your heart is full of fertile seeds, waiting to sprout.

MORIHEI UESHIBA
Japanese martial artist

*We have neglected the truth that a good farmer is
a craftsman of the highest order, a kind of artist.*

WENDELL BERRY
Poet, novelist, environmentalist and farmer

1
Meet My Friend Viswa
The Leader Who will Show You the Way

The task of the leader is to get his people from where they are to where they have not been.
—Henry Kissinger
Former National Security Adviser and former Secretary of State of the US

Viswa was my classmate in postgraduation, and we studied Chemistry together. He was undoubtedly a brilliant student, certainly better than me. He had diverse interests—music, theatre, painting and technology. He came from a village called Agaram, near Tirunelveli in Tamil Nadu, in the southern part of India. His family owned a large tract of land there and was involved in agriculture. He lived in a joint family.

Years later, we had the opportunity to work together in the fertilizer manufacturing industry. The factory was near Chennai, and we worked there for two years. Although we were in different units—I was in the urea plant and he was in the NPK plant—we would occasionally meet during the same shift.

One night, during my shift, I was sitting on the second floor of my unit. It was around two o'clock in the morning. Viswa had a half-an-hour break and came to see me. "Hey, what are you doing sitting quietly?" he asked. He grabbed the paper I had in my hand, read it, and then stared at me.

"Writing poetry? That too, at midnight…," Viswa remarked.

I was quiet for a moment, then smiled.

"This is my first love—poetry," I replied. "I enjoy writing. It brings me so much happiness and fulfilment. The compressors and pumps are pounding in my head every minute, but I love to sit and gaze at those twinkling stars, the silvery planets and the vast expanse of the heavens. What a masterpiece by the Creator! Every heavenly body is in its own place in the galaxy, yet they all look at each other and work together. There's so much order and beauty in nature."

Viswa looked deeply into my eyes and said, "Let me tell you, Bala, you're in the wrong place. Industry isn't where you belong. You need to do something you love; otherwise, over time, it will become a burden, and you'll end up mentally drained. There has to be an alignment between your aptitude, passion and work." His words left me in deep thought.

A few months later, I resigned from my job after a disagreement with my boss. I didn't see Viswa after that. Then, a month later, a terrible accident occurred at that factory, and nearly half a dozen of my former colleagues lost their lives. I had spoken to Viswa around that time, worrying about his safety.

Years went by. A few months ago, while I was on a train to Bengaluru, I heard a familiar voice: "Hey Bala, how're you?" It was Viswa. We chatted for nearly an hour on worldly things.

When I asked him what he was doing, he simply smiled and said, "I'm doing what I love. I'm following my passion—farming."

Before we parted ways, Viswa promised to visit me soon.

True to his word, two weeks later, Viswa arrived at my house. "Pack your things. We're heading out for a few days," he said. He drove me to a nearby village—a peaceful and tranquil place. The village seemed to have only two or three streets. As we drove through, people from nearby houses looked our way, and I heard someone shout, "Viswa sir has arrived!" The voice carried a note of joy.

We stopped in front of a relatively big house on the village road. As he walked towards his house, Viswa bent down, scooped up a handful of soil and applied it to his head. "This is my God. It's divine," he laughed.

I didn't respond. I simply saw him in a new light.

We entered his house, and he continued, "That accident changed my life. I quit the job and joined a corporate company in their HR department, where I worked for five years. It was a great learning experience—dealing with people and their issues, listening to their joys, aspirations, disappointments, sorrows and failures. Good package. Comfortable life. But I wasn't happy with the fast-paced city life. I stayed single. I thought I had a different calling in life. So, I came here and started farming. Now, I love what I do. Look at the coconuts, mangoes, plantains, paddy, pulses… they're my family. I talk to them. I gossip with them. I laugh with them. They have real lives… no one understands."

He looked at me. Viswa appeared very contented and peaceful. I smiled.

On the wall, there was a wallpaper with a quote: "*It feels good at the end of the day to know you made a product that other people are going to enjoy.*" – Jericho Sanchez.

I stood there, reading it over and over.

"Do you like that quote?" Viswa asked me.

I nodded. "It has a deep meaning."

Viswa smiled. "You're right. Farming brings a happiness that no other profession can offer. You see life growing, and you feel like a creator. Of course, it has its share of problems. It lands you in unpredictable situations. But the joy comes only when one cultivates."

"By the way, Bala, what's the focus of your training programmes?" he asked, interrupting my thoughts.

"I conduct a lot of programmes on leadership development, lifelong learning, team building, collaboration, life skills …"

"Oh … you're doing a different kind of farming…," he said with a smile.

I fell silent as Viswa's words stirred something deep within me, resonating with my conscience. *Could leadership development be a different kind of farming?*

To be honest, I couldn't sleep that night. I wondered if Viswa had managed to.

The next morning, I admitted to Viswa, "Cultivating a human mind is no easy task."

He responded thoughtfully, "Think of how a mother nurtures a foetus in her womb. The foetus requires not just medical care but also emotional, psychological and personal attention. Every single cell of the foetus needs love, care and compassion. It's like a tiny seed that will eventually grow into a fruit-bearing tree within a social setup."

Viswa was right. I couldn't disagree with him.

"Leadership and farming?" I smiled, murmuring to myself.

Viswa heard me and asked, "Aren't you a leader?"

"Yes," I replied.

"Who said you are a leader?" he continued.

The question hit me hard. I stood there, silent and baffled.

"Who are you leading?" he asked.

I preferred to remain quiet.

"That's the problem," Viswa said. "Most of us aren't aware that we must first lead ourselves before we lead others. The entire Indian philosophy is focused on developing that self-leadership."

I listened patiently.

"Bala, my considered view is this: if one cannot lead onself effectively and purposefully, they cannot lead others," Viswa added.

I thought about it; he was right. *Had I truly been leading myself all these years, or merely drifting along with the flow?* I slipped into an introspective mood.

During the discussion, Viswa explained that many of us lead ourselves without even realising it. "A conscious, thoughtful and programmed effort in self-leadership," he said, "can bring better rewards in a shorter time."

"To which school do I go?" I asked, curious.

Viswa smiled and replied, "School? Family is the school, and parents are the teachers. The community is the school, and the senior citizens are its great teachers. Life is the best school, which provides you with experiential learning."

"What about the branded institutions that claim to teach this?" I interrupted.

"Well, most of them teach you management—principles, strategies and challenges. And of course, a few lessons on leadership too," Viswa said assertively. "And there, you learn some basic principles about leadership through texts, lectures, projects, case studies, and short-term engagements with workstations. But real leadership is much more than that. You only learn it through life."

"Don't you think that's a sweeping statement? Don't they prepare you for life?" I asked.

"Certainly, they do," Viswa acknowledged. "But they prepare you for a degree. There is no guarantee that you are going to learn. Learning is a personal choice; it's not something delivered top-down."

"I'm making a conscious assertion," he continued. "In institutions, you understand leadership through a third-party perspective. They provide you with the experiences of several proclaimed leaders, perceptions that are diverse, contextual and reflective. But in life, it's about self-learning."

"What do I do to know if I'm emerging as a leader?" I asked Viswa.

"Do you give yourself time to reflect on your own self?" he asked.

I remained silent, and my silence conveyed exactly what I wanted to say.

"We spend so much time talking to others every day, with or without purpose. Sometimes we talk aimlessly, and it takes us nowhere. We talk to everyone except our own selves," he said emphatically. "We can plant a seed in the right soil. We can facilitate its growth by providing the necessary light, water and manure, but the seed must grow on its own. So must we."

"I agree." I recalled the words of Mark Twain: *A man cannot be comfortable without his own approval.*

I responded, "Do you think that when you provide all these inputs and resources, growth is guaranteed?"

Viswa smiled. "It depends on the inner content of the seed. In the same way, it depends on your inner being. Unless you have the right attitude, a positive mindset and a willingness to grow, a lot of inner engineering is needed."

"If someone grows, is there any guarantee that they'll become a leader? Is there a connection between growth and leadership?" I countered.

"Certainly not. But if you lack the right inner content, even if others declare you a leader, you may not sustain that leadership for long," Viswa answered.

I remembered the words of one of my mentors, who often said, *Bala, always keep in mind: 'Deserve and Desire'.* I shared this with Viswa.

"True," he said. "Enriching and empowering the self are the basic gateways to leadership. I recall the words of the famous poet Rumi, which have immense meaning: 'You are not a drop in the ocean, but you are the entire ocean in a drop'. That's the direction in which we need to empower people."

I joked, "Do the seeds on the farm also do that?"

"Of course," he responded quickly. "The entire tree is in the seed. For those who can visualize the power of the tree within the seed, the essence of leadership becomes clear."

"But seeds don't grow instantly. Plants assimilate energy, gaining strength and growing in a systemic manner. They aren't like humans, who are often greedy for overnight growth. They don't look for a mentor who will wave a magic wand to make them grow in a split second," he added.

We both laughed.

"So, you rule out the possibility of 'instant leadership', like 'instant coffee'?" I asked.

"Well, there can be instant leaders, granted that status by a vote, a law or by dynasty. There can also be accidental leaders who emerge in specific situations. In such cases, some facets of leadership are briefly showcased, but they rarely stand the test of time," Viswa said.

He continued, "Even among seeds and plants, growth doesn't happen within a set timeframe or according to any external prescription. They take their own sweet time."

"Bala, it's time to move on …," he chuckled.

And with that, we decided to move on to other engagements.

REFLECT

1. Leadership is a continuous process of growth.
2. Self-learning is essential to individual development.
3. Self-directed learning fosters purposeful, lifelong growth.
4. Leading oneself is a prerequisite to leading others.
5. A positive mindset acts as a catalyst for self-leadership.

YOUR LEADERSHIP QUESTIONS

1. Have you ever identified your passion and considered how you want to pursue it?

2. Why do you want to become a leader, and what is the primary objective you hope to achieve on your leadership journey?

3. How often do you invest in your personal growth?

4. Have you ever reflected on the direction you want your growth to take?

5. What kinds of inputs do you believe are essential to enrich yourself and grow beyond who you are today?

2
Nurturing the Seed and the Self

Because you are alive, everything is possible.
—Thich Nhat Hanh
Vietnamese spiritual leader, poet and peace activist

"Are you ready for coffee?" Viswa knocked on the door and asked. I opened it. "Come along. I've made some nice filter coffee," he said. The aroma of the coffee was gravitating. He handed me the coffee in a brass tumbler and a *dabra* (a small, circular, deep vessel akin to a saucer). The coffee was piping hot, with a creamy layer of *nurai* (the Tamil word for foam) on the surface.

"Take the first sip, and it should transport you to an entirely different world," Viswa smiled. And it did.

He continued, "I don't need a large cup. A small one is enough, but it should be of quality. The same goes for anything we do—whether big or small—it need not speak volumes, but it should reflect the essence of what we intend to accomplish."

I raised my eyebrows.

"You know Bala, that's something I've learnt from farming. You may have several acres of land, but if there's no produce, what's the use? It's important to see how much you can yield from every inch of that land. As Paul J. Meyer, a pioneer in self-improvement, said, 'Productivity is never an accident. It is always the result of a commitment to excellence, intelligent planning, and focused effort'."

I fell silent, reflecting on how Viswa managed to apply management principles to his farming practices.

"And you need to go one step further—apply the principles of TQM to agricultural farming," he added.

"You mean the principles of TQM (total quality management) by Edwards Deming?" I asked.

"Yes! Edwards Deming, whose principles are followed by several others from different countries," Viswa began. "Japan's post-World War II recovery owes a lot to Deming, the architect of the quality movement. I learnt how his principles influenced not only large-scale strategies but also small, practical actions. It was indeed a great learning that brought transformation. Deming once said, 'If you don't understand how to run an efficient operation, new machinery will only bring new problems in operation and maintenance. The surest way to increase productivity is to better administrate both man and machine.'"

He sighed, "Many of us believe that changing the machines and tools will solve all problems. They won't. Maybe it can improve efficiency, but unless you improve your proficiency, the machines don't matter".

"And what about farming?" I asked.

"Farming is one activity where all your management principles come into play. Beyond that, it helps one evolve as a seasoned leader. You need a lot of wisdom to be an effective farmer. There's an interesting statement by American author Kelsey Timmerman about a farmer's life: 'Farming isn't something that can be taught. Each plant tells its own story, which has to be read repeatedly,'" Viswa said.

"Is it important to know how much we produce with every inch of the land?" I asked, and then added, "But it's one man's job, unlike in a corporate setting, a business or an industry. In larger organisations, you not only evolve as a leader but also develop other leaders. How does this apply to farming?"

Viswa turned to me and said, "Sorry, but you're mistaken. In farming, everything you do, right from identifying the soil to cleaning up the land after harvesting and preparing it for the next crop, is a lesson in leadership. It's exactly like developing leaders. And there's a huge opportunity to create a community of leaders. To be brutally honest, it's even more difficult. Masanobu Fukuok, the Japanese philosopher and farmer, once said: 'The ultimate goal of farming is not the growing of crops, but the cultivation and perfection of human beings.'"

I listened carefully. His words felt authentic. He wasn't just saying things because he had to.

"How's the coffee? Has it gone cold?" Viswa interrupted my thoughts.

"Not much," I replied.

"That's the problem. When you're enjoying a conversation, you miss the aroma and taste of the coffee. In corporate life, we face the same issue: we get so engrossed in the mundane things of our achievements that we miss life itself. We miss the work-life balance. While we're focused on work, we forget to live. In farming, we celebrate both life and work. We enjoy our work as much as we enjoy our lives," Viswa said.

"Viswa, when do you usually wake up in the morning?" I changed the direction of the conversation.

"Three o'clock. Some days, half-past three," he said, looking at me.

I was shocked. "Three in the morning? What do you do so early?" I asked.

"It's not early for many. In the Bhartiya tradition, it's called *Brahma Muhurtha*. It's supposed to be the time when the mind is most active," Viswa explained.

"Can you elaborate a little more on that?" I asked.

"In our ancient traditions and astronomy, a *muhurta* is a sacred unit of time lasting 48 minutes. There are 15 *muhurta*s during the night. *Brahma Muhurta* is the penultimate *muhurta* that occurs just before sunrise. It's considered the most auspicious time for practicing yoga, meditation and other spiritual disciplines. Action taken during this time is believed to relieve stress, anxiety and other negative influences. Waking up at this time helps me in several ways. It greatly enhances my focus throughout the day, triggers positive energy and allows me to work with greater peace. It also provides me with additional time," Viswa said.

"Sounds great," I said.

He continued, "It's considered the best time for reading or engaging in any creative thinking. Attention levels are much higher during this time."

"How does it help a farmer?" I asked.

"There's an interesting saying by Edgar Watson Howe, the famous American novelist: 'Even if a farmer intends to loaf, he gets up in time to get an early start.' Many farmers wake up early, long before sunrise. They prepare themselves for the day during this time and set off as soon as the sun rises. Their energy levels are high, they're calmer and more peaceful, they've better clarity, and their work quality is best. They're more focused on executing their tasks. It's said, 'During *Brahma Muhurta*, the universe whispers its secrets to those who are ready to listen.' And that's a powerful message," Viswa said.

I sat there like a student, inspired by the teacher. I smiled.

"Why are you smiling?" he noticed.

"Many of my corporate friends go to sleep only at that time," I replied.

"Yes, in many cases, they put time into their work, not their work into time. They continuously raise the bar for their performance, but lower the bar for themselves. They struggle with the 'standards of life' at the expense of the 'quality of life'," he added.

"Not necessarily," I countered.

"Well, Bala, I'm not discrediting them. But in many cases, they suffocate in their work. They experience a lot of stress. They're in a constant battle with themselves and their jobs," Viswa said.

"Work, numbers, targets, profits …," he lamented.

"I think it has a significant impact on their personal lives too," I said.

"In farming, there's certainly more peace. You learn to understand what's possible and what's not," Viswa said.

"Therefore, do you think corporates and businesses target impossibilities?" I asked.

"No. They enjoy difficult achievements and feel happy accomplishing them. But every achievement becomes the base for the next one. The cycle goes on … and no one understands where it starts and where it ends. It's a rat race. The definition and meaning of growth become truncated. Many suffocate in life, chasing the myth of success," Viswa said.

"Building fortunes? Making more money?" I asked.

"Of course. But is money truly wealth? Or is wealth different from money?" he asked.

"Wealth is a holistic concept, and money is just one subset of it. Health, knowledge, skills, talents, family, relationships, trust—all these are components of wealth. One shouldn't sacrifice everything else just to acquire money. Of course, money is an important ingredient to achieve many other things in life," he added.

I narrated an interesting episode to Viswa.

"Decades ago, I was invited to a corporate conference for HR leaders. During one of the sessions, there was an interesting group discussion, and I was asked to be the rapporteur. When the topic was assigned to me, I was shocked. It was titled 'After Forty, What?' I didn't quite understand it at first. Later, it was explained to me: 'Many people enter the corporate universe in their mid-20s. They work tirelessly, day in and day out—or rather 24x7—in air-conditioned offices. They don't even know whether it's raining or shining outside. They get married but have no time for their families. By the time they're 40, with huge sums of money in their bank accounts, they start wondering, 'Have I missed something in life?' And that's when they realise they've missed their entire life. They then take a golden handshake from the organisation. And many of them start wondering, 'After Forty, What?'"

Viswa laughed. "Is the upper limit of 40 coming down now?" he asked, satirically.

He continued, "Leisure, relaxation and sleep are essential for both physical and mental health. If you don't sleep well, your anxiety rises, and your stress and tension increase. You also need to spend quality time with your family. Leadership is a holistic concept; it doesn't exclude one's family. That's why I said 'self-leadership' is the core of all other types of leadership."

I replied, "It's a survival problem for many. Their minds work like factories late into night, producing ideas, plans and strategies. They call it a numbers game. Their growth appears to be defined by the numbers they achieve."

"And in many cases, these numbers are manipulated to show achievement levels—they're more apparent than real," I added.

"Well, excuse me for some time," Viswa said. "It's time for me to complete some daily routines. Please bear with me." He walked away.

I quietly followed him. He picked up a broom and began sweeping the house entrance. After spraying water on the floor, he moved around the courtyard, sweeping up fallen leaves and gathering flowers from the blooming plants.

Viswa noticed me watching him and said, "This is jasmine. Quite fragrant. But there are several varieties of jasmine, and not all of them are fragrant. You need to nourish them just like you would nourish leaders in your organisation. By the way, leaders of any organisation are the ones who carry the fragrance of its goodness. They're like the ambassadors of the organisation."

I was spellbound!

"To nurture such people, you need to follow the farming practices. You need to study the quality of the soil; understand what kind of seeds will grow in each soil. Similarly, you need to recognise that not everyone displays leadership in every situation. Just like different plants grow in different soils, different leadership styles grow in different organisations," he added.

He continued, "I was reading *Empowered: Ordinary People, Extraordinary Products* by Marty Cagan and Chris Jones and came across this sentence that really stuck with me: 'Leadership is about recognizing that there is a greatness in everyone, and your job is to create an environment where that greatness can emerge'."

"Yes, that makes sense," I said.

"That's good enough for the morning," Viswa replied. "Have your bath and get ready for breakfast. Don't expect anything fancy—just fruits, salad, some cooked millets, and some tea—ginger, cardamom, cinnamon or whatever you prefer. But as for me, I only drink hibiscus tea."

I was quite surprised by his way of living and said, "You'd fit into the category of a transformational leader."

"Leader?" he glanced back. "I don't lead. I just live. If anyone considers my lifestyle and the way I work as worth emulating, they're welcome to do so—that's not my concern. I'm not in the rat race for leadership. Then you start expecting recognition, celebrations, and awards to display in your office. I'm out of all that. I just enjoy what I do and keep moving forward."

He continued, "You know, the great Chinese philosopher Lao Tzu once said, 'A leader is best when people barely know he exists, when his work is done, his aim fulfilled, they will say: we did it ourselves.'"

"Do you think there's substance in his argument?" I asked.

"Indeed, there is substance if we read between the lines," Viswa responded.

I recalled a statement from one of the articles I had read online about "farming responsibility". It said: *Why do farmers farm, given their economic adversities on top of the many frustrations and difficulties normal to farming? And always the answer is: 'Love. They must do it for love.' Farmers farm for the love of farming. They love to watch and nurture the growth of plants. They see them smiling, they understand their passion, see their tears, their dance, their romance and they see the plants living all the time. They love the sunshine, the rains, the breeze, the thunder and the weather, even when it makes them miserable.*

Certainly, Viswa is a great farmer. And of course, a leader too!

REFLECT

1. Concern for quality is a key attribute in leading any process towards growth.
2. Effective leaders put work into time, not time into work.
3. Leaders understand the difference between wealth and money.
4. Leadership involves creating and nurturing social wealth.
5. Leaders who lead by example naturally attract followers.

YOUR LEADERSHIP QUESTIONS

1. Do you think each morning offers a new page in your pursuit of leadership? What would you like to write on that page today?

2. How do you define quality in leadership? What steps can you take to enhance quality in specific areas of your leadership?

3. How do you plan to maintain a healthy work-life balance? What are a few things you would like to do to achieve this?

4. What qualities or strengths in your profile might naturally draw others to your leadership?

5. Which aspects of your leadership could inspire others to see the leader in you?

3
Mentoring the Farm and the Leader

Soil is a living ecosystem and is a farmer's most precious asset. A farmer's productive capacity is directly related to the health of his or her soil.
—**Howard Warren Buffett**
American philanthropist, political scientist and writer

As I walked towards the bathroom for a shower, I saw a wall hanger with the statement: "*Water is precious; Save Water*". I was a bit surprised because there didn't seem to be any scarcity of water, especially since Viswa's farm was located in fertile agricultural land. I assumed the water table here was quite good.

At the breakfast table, I brought it up with Viswa. He responded with a smile. "You're right, there's no shortage of water here. The water table is indeed good. But that doesn't mean we should use it carelessly. Water is a valuable resource; it's the life force. There are so many places where people struggle just for a single bucket of water. This is something I learnt during my corporate days. Resource management, process management and waste management – these are three critical aspects of any productive system. A leader who isn't sensitive to these may not succeed in their career."

He continued, "Back here in the village, people are highly aware of how to use natural resources wisely. When I started talking to them about the global importance of water and other resources, and how the world is becoming increasingly concerned about sustainability, they were eager to contribute. They wanted to help educate the next generation."

Viswa paused, then added, "They've also realised careful use of resources has helped reduce operating costs in some areas. They're responding positively to concerns like the 'energy crisis', and I truly believe they could set a model for other villages. They're on an impressive learning curve."

I was convinced by his point of view. What impressed me most was his commitment to developing leadership among people. "It is truly a great service, Viswa," I said. "I think I'd call it 'Leadership Farming'. Nurturing leadership is like farming. It requires great sensitivity."

He burst out laughing and said, "You said it right, Bala. Nurturing and developing are the core elements of farming, and as you rightly pointed out, it demands sensitivity. If we look at leadership development through the lens of farming, we'll find many similarities. On the field, we nurture crops and harvest the produce. On the social front, we nurture human minds and harvest productivity for the nation. Isn't that true?"

He managed to provoke my thoughts once again.

He had laid out some plates, teacups and a jar filled with warm water on the table. "Feel free to serve yourself. No formalities here," he smiled. His tone was warm, filled with genuine camaraderie. "The only thing I tell people is to take as much food as they need, but don't waste it. I've seen people serve themselves heaping portions and just leave it uneaten. They don't realise how much labour goes into producing each grain. A 2023 United Nations report states that the average food wastage per person worldwide is about 79 kg annually. That translates to nearly a billion meals wasted every year."

He heaved a sigh and continued, "Listen, it's all about attitude. And let me add this – 'Consume only what you can digest.' Overeating is dangerous; it's a hazard. But today, we've become a consumerist society. We buy and consume things regardless of whether we need them, want them, or even know how to use them. This creates unwarranted competition in the market and drives up costs."

I listened patiently, feeling like a student in a classroom with Viswa as the teacher. His words reminded me of my grandfather, who used to quote the Sanskrit saying,: *"Annam Brahma"*, meaning "Food is God".

Viswa continued, "Consumerism has distorted our lifestyles in many ways. As American musician Aloe Black puts it, 'We live in an era of consumerism and it's all about desire-based consumerism and has nothing to do with things we actually need.' Many families buy things they don't need. Some feel compelled to have a television in every room or a separate car for each family member."

"But several corporates define success by creating false consumerist needs among people—either through fear or insecurity—and then aggressively push their products," I said.

"You're right, sir. This even happens in the medical field, where a false trauma is created in the minds of people," said Amar, a member of the team of young agricultural graduates that included Shyam and Ankita. He went on, "I came across two books in the library titled *Emotion Marketing: The Hallmark Way of Winning Customers for Life* by Scott Robinette, Claire Brand and Vicki Lenz, and *The New Science of Customer Emotions*, published by Harvard Business Review and now considered a classic."

"Yes, many advertisements are provocative. They manipulate people's emotions, creating needs or fuelling greed," I said.

Viswa laughed and said, "W. Edwards Deming, American statistician, educator, and consultant, once said, 'The aim of leadership should be to improve the performance of man and machine, to improve quality, to increase output, and simultaneously to bring pride of workmanship to people'. I think, in that sense, we need to learn from the villagers. They embrace 'minimalism'."

"I think the concept of minimalism was first deeply practised in Japan," I said.

"Correct," Viswa nodded. "In Japan, they have limited space, so they must manage their space and needs accordingly. Minimalism becomes a pathway to effective fiscal and resource management. But don't forget, minimalism was also strongly recommended in Indian thought culture. There's an old Tamil adage, '*Siruga katti peruga vaazh*', which means, 'Live gloriously by saving even small things'."

Amar intervened, "Sir, I once read a beautiful quote by William Rogers, an American diplomat: 'Too many people spend money they haven't earned, to buy things they don't want, to impress people they don't like.' Isn't that true?"

Viswa nodded thoughtfully. "Yes, it is. I think leaders in industries are slowly beginning to learn these lessons. One trend

I've observed recently in many business houses is how carefully they now manage their inventories. There's a growing awareness around stockholding and the strategic use of inventory. Supply chain management has become a critical discipline. It's built on planning and offers multiple advantages."

He continued, "Inventory management in agriculture, however, requires a lot of prudence, especially because it deals with both perishable and non-perishable products. The logistics for managing these two types are vastly different."

Viswa paused before adding, "Leadership in agriculture, therefore, needs a wide spectrum of skills—more precise, more contextual, more empathetic and more pragmatic than in other types of professional leaderships."

He then turned to me when I queried, "By the way, my friend, I have one more question. Inside the room, I saw a quote near the mirror …." Before I could complete my question, he responded, "Yes. It reads: *'The best leader who can lead you to success is standing before your mirror; it is none except yourself'*."

"Indeed, it is a profound statement," I said.

"Not only profound, but it is the truth," Viswa replied. "I've gone through several learning experiences in life, and what I've learnt is that, at the end of the day, you're your own leader. If you've no commitment to yourself, you'll never be able to lead."

I agreed with him. "That reminds me of the famous words of the Greek philosopher Aristotle: 'Knowing yourself is the beginning of all wisdom'."

Viswa nodded. "You're right, Bala. Before you aspire to become a leader in whichever field you want to be, you must first know your strengths and weaknesses. You need to know how much knowledge you possess in that area and what else you need to learn. It's also important to understand the social dynamics related to that field. You must assess whether you have merely a desire to pursue it or if you truly have the capability. Many people walk into certain fields driven only by desire, but without the basic inputs regarding the area they are venturing into."

I said, "Yes, I know several people who venture into professions or businesses simply because they're gravitated by the success of others. They leap in without understanding the ground realities, and more often than not, it ends in disaster."

"Making money should not be the singular focus of leadership," Viswa said thoughtfully. "If that's your only aim, you might be a businessman, but not necessarily a leader. Not everyone who makes money is a leader. Some shine for a time and disappear, while others appear like a flash and then fade away. Others spend huge amounts just to be seen or branded as leader. It becomes more of a business or a trade. If one wants to nurture leadership, they must focus on generating wealth."

"Wealth, not money?" I asked, listening attentively.

"Exactly," Viswa responded. "Money is just one small component of the wealth you generate. There are several other elements—health, knowledge, skills, competencies, trust, family and much more. Some of these might not have an immediate market value, but they carry immense intrinsic worth. To me, health comes first—physical, mental, emotional, psychological and spiritual. Without strong health, you won't last long in the leadership journey. Making 'quick money' or achieving 'overnight successes' should never be the goal. That's why I say self-awareness is critical to leadership."

I looked at him and asked, "Are you leading yourself?"

He laughed. "I am trying to."

He paused for a moment, reflecting. "Leadership is a journey, not a destination. To remain a leader, you must keep learning, keep discovering yourself and stay aware of your social and professional equity within the community you're part of. If you stop doing that, you become stagnant. And when you're stagnant, you're bound to face repeated failure as others continue to grow."

"And do you have any such opportunity to gain experience in farming?" I asked.

"OMG! What a question!" Viswa exclaimed. "One can't be a farmer unless one learns something new with every dawn. Every ray of sun brings a fresh challenge to the farmer."

"Sunrays?" I asked.

"Yes, both their presence and absence. You have to keep your eyes and ears open to everything happening around you. An intense sunray can be as threatening as a bolt of lightning or a crack of thunder. A farmer is finely attuned to every sound made by the cattle and animals. He listens to their needs, their pain and their joys. He hears the rustle of leaves, the chirping of birds and the rhythm of the wind. He understands these signals and responds accordingly. He knows what each one means and what action to take," Viswa responded.

"Are you saying the farmer lives in constant fear?" I asked.

"No. Certainly not," Viswa said firmly. "He lives in a state of a constant learning, a continuous sense of uncertainty. He has no option but to be a perpetual learner. And above all, he must have sustainable courage. A farmer needs to be smart, clever, sensible, sensitive and wise."

"You seem to be describing the ideal qualities more than what an ordinary leader might need," I interrupted.

"True," Viswa nodded. "Farming is, in fact, a focused exercise in leadership. It encompasses many forms of leadership: pragmatic, strategic, servant, participative, and of course, thought leadership…."

"Hold on," I said, surprised. "Are you saying a farmer can be a model for a leader?"

Viswa responded, "Absolutely. It's no joke. This is real; it's an exercise in living leadership, not just a theoretical concept. Unfortunately, no book teaches it. Sadly, no book teaches it. And no book ever will. Our ancestors lived it. As Wendell Berry, the environmental activist, once said, 'We have neglected the truth that a good farmer is a craftsman of the highest order, a kind of artist'.'

"Are they?" I asked, intrigued.

"Without a doubt. The very foundation of social order across centuries was built on farming. It was only with the rise of industrial society that the paradigm shifted—from self-leadership to what I call programmed leadership," he explained.

I laughed, "Programmed leadership? You mean…?"

"Yes," he said. "We began to programme leadership concepts to fit institutional needs, focused on controlling resources, managing people and projecting one-upmanship, as if leadership were a game of power. But true leadership isn't about power. It's about making a meaningful contribution to the time and space you live in."

"So, where did you get your first lesson in leadership? From farming?" I asked, genuinely curious.

"From the soil," he laughed. "You'll understand when I take you to my farms. As Henri Alain, a professor from Harvard and political economist, puts it, 'Life on a farm is a school of patience; you can't hurry the crops or make an ox in two days.' But for now, the fruits and salad are ready to give you the energy you need for the day."

The breakfast table was filled with not just food but also with knowledge, skills, and wisdom.

> **REFLECT**
>
> 1. Farming skills and leadership skills have much in common.
> 2. Consumerism is not the hallmark of growth.
> 3. Self-awareness is critical to leadership.
> 4. Leadership is not about competition or winning a race; it is about pursuing path-breaking activities.
> 5. Leadership is not about one-upmanship or asserting identity; it is about continuously creating a new version of oneself.

YOUR LEADERSHIP QUESTIONS

1. What are the three most important things you think are needed to nurture the self?

2. What are three qualities you admire in the leaders around you that you would like to learn and follow?

3. How do you think you can demonstrate your leadership traits without competing with anyone?

4. What do you think are two unique qualities or identities you possess that can support your leadership journey?

5. What are two practices you would like to adopt to better understand and master yourself?

4
The Growing Seed and the Growing Leader

Leadership and learning are indispensable to each other.
—John F. Kennedy
Former President of the United States

I was fully dressed and eager, ready for the field visit. As Viswa came out, he stared at me for a moment and suddenly burst into hearty laughter.

"What's so funny?" I asked.

"To the farmers in the field, you'd look like a magician," he chuckled. "Are you really going on a field visit in a full-sleeve shirt and trousers? A formal wear? You've only missed the tie or maybe a bow tie! Come on Bala, you need proper attire for this! Change into a *lungi* or a *dhoti* (both mean a draped cloth used by men as casual attire). This isn't a conference or a training programme. You're going to experience farming. You need to be relevant to the place and the work you're stepping into. The only place your outfit might get attention is at a fashion show or a film festival."

His comment reminded me of the famous costume designer Edith Head's words: *You can have anything you want in life if you dress for it.*

I stood silently, looking at Viswa carefully enough to remember what I was supposed to wear. He was dressed in a blue T-shirt and a simple *dhoti*.

"Come on, hurry up," he said patiently. "I'll wait for you."

A few minutes later, I returned, dressed appropriately. I was relieved when he said, "That looks great. Much more appropriate. Let me tell you something: a leader needs to be sensitive to their environment. Otherwise, they end up like a piece of material for window shopping. "

He continued, "For example, take a foreman or a product manager in a factory. They should wear what the workers wear. A marketing manager should dress like the other representatives. That creates a sense of ownership and participation. Makes sense, right?"

"Ownership?" I raised my eyebrows.

"Yes," Viswa said. "When you step into a role or a workplace, you must own what you do. Without ownership, you appear alien. You keep a distance from the team. And leadership doesn't work at a distance."

He laughed, and the two young boys and the girl in her early 20s sitting nearby laughed along with him. "By the way, I forgot to introduce my young friends," Viswa said. "You've already met Amar. The other two are Shyam and Ankita. All three of them are agricultural graduates from this area. Initially, they were looking for jobs in corporates or banks. Then someone told them about what I do and they decided to join me."

Amar smiled, "That was the turning point in our lives. Viswa sir helped us realise our strengths and showed us that we have a future as agriculturalists and entrepreneurs."

Shyam added, "He introduced us to someone we had never met before … our own selves!"

"He taught us 'Leadership Farming', or rather, leadership through farming," Ankita said. "Each of us has a small piece of land inherited from our parents, and we've started growing vegetables there."

"Shyam and Amar are also exploring floriculture and export," Viswa explained. "They've done their research and found global opportunities in flower exports. They already have good-quality produce that's in demand locally. They're applying their academic knowledge and skills directly in the field."

He paused, then continued, "To put it bluntly, they had the aptitude to learn. They're excellent self-learners. All I did was facilitate that process. You can't force anyone to learn if they lack the willingness or ability to absorb information. As the saying goes, 'Learning cannot be caused'."

"He lit the fire in our bellies, if I may say so," Ankita said, smiling. Everyone laughed.

"And it's been burning ever since," added Amar.

"He's a great mentor," said Ankita sincerely.

Viswa raised an eyebrow in mock surprise. "Oh my God! You've never called me that before, Ankita. I'm actually a little scared of that word. It's easier to be a leader than a mentor. A mentor's responsibilities are many and quite difficult. Talking about mentors, Steven Spielberg, the famous American film director, once said, 'The delicate balance of mentoring someone is not creating them in your own image, but giving them the opportunity to create themselves'."

Ankita listened quietly, her expression thoughtful. I couldn't help but notice how well she spoke English.

"You know, she has a real flair for English literature," Viswa commented.

I was pleasantly surprised. "A village girl pursuing a degree in agriculture and with an interest in English literature?"

"Sir, not just English literature, but Tamil literature too," Ankita laughed. "Literature has a humanising effect."

Viswa said, "Bala, did you know that many of our greatest literary geniuses were farmers? Anyone who thinks that great writers are only produced in colleges or universities is completely mistaken. A talented cook doesn't necessarily need a degree in culinary arts, and a passionate musician doesn't need a degree in music. Many of them simply live and learn through their passions."

I nodded, "Yes, we see such people among farmers, artisans, textile workers, artists and the like. For many of them, it's either a deep passion or a spiritual pursuit. But do you think that's because agriculture was once the dominant profession?"

"Well," Viswa said, "regardless of the profession, if you have an inner drive to create or express something, it will find its way out. Every human being wants to communicate who they are, whether through poetry, stories, theatre, dance, painting or any other form. Educating people is simply about letting them manifest fully."

I was impressed. "That reminds me of a quote by Swami Vivekananda: 'Education is the manifestation of the perfection already in man'. So, Viswa, you're an educator too, aren't you?"

He smiled. "Each of us educates others in some way. But the real question is how many of us are willing to learn from others? A good leader seeks opportunities to learn, no matter the source."

His words were indeed worth reflecting on. "These young people," Viswa continued, gesturing toward the group, "want to become agricultural entrepreneurs."

"Sir, Peter Drucker, the famous management guru, said, 'The best way to predict the future is to create it,' and we're trying to live by that statement," Ankita responded.

"*Leadership and entrepreneurship—how are they related?*" I murmured to myself.

Viswa, quick to catch it, responded with a smile. "I'll tell you. Every leader must be an entrepreneur—an entrepreneur of ideas. And every entrepreneur should be a leader. A leader who is constantly learning, curious and discovering, so they don't lose their entrepreneurial spark. These young people are both leaders and entrepreneurs. Talk to them and you'll see for yourself."

"We are like seeds sown in the right soil, or rather, fertile soil," Shyam chimed in. "We're just beginning to explore, and the support and nurturing come from Viswa sir."

I turned to Viswa, intrigued. "Who are you, really? A leader, a businessperson, a mentor, an entrepreneur, a preacher … a teacher?"

He dismissed the labels with a gentle smile. "None of those. I'm just an average human being, trying to remain aware—of myself, of the purpose of my existence, a seeker of growth, attempting to contribute a little to the people and things around me. In short, a continuous and willing learner."

"And that makes you a perfect leader," Ankita said warmly. "Sir, can we call you a humanistic leader?"

"Every leader is a humanistic leader in one way or another," Viswa responded. Then he added, "There's a quote often attributed to George Eliot, the famous poet: 'Wear a smile and have friends; wear a scowl and have wrinkles. What do we live for if not to make the world less difficult for each other?'"

"Enough of this discussion. Time to get moving... Did you all have breakfast?" Viswa asked.

"Yes. I had *ragi* porridge," said Amar.

"I had some papaya, watermelon and two *idli*s," said Shyam with a smile.

"I had a glass of milk," added Ankita.

"Only milk?" Viswa asked.

"Yes, sir. I was getting late," she replied.

"Wait. Have these two bananas. You can't work without energy. And honestly, you shouldn't." He gave her the fruits.

I was closely observing how he interacted with them. "Viswa, don't you think these young people can take care of themselves?" I asked.

"Of course they can. But as a team leader, I need to be conscious of their well-being. You can't lead a team unless you understand them inside out. While you don't have to get personal, you must be sensitive to their ecosystem," Viswa replied.

"How does that help you?" I asked.

"In any organisation or system, the emotional content of the team members is crucial. I prefer to call them members, not employees. If someone carries emotional pain whether in the mind, attitude or spirit, they won't be able to discharge their work fully. Pain drains energy. Any work, something like farming, needs joy and wholehearted involvement. And when you're dealing with seeds and crops, bringing in pain or negativity affects the process. The plants feel it. They absorb the energy around them. That energy, believe it or not, can even influence the people who consume those crops later," Viswa said.

He paused for a moment, then continued, "There have been multiple studies showing how plants respond to their ecosystem. One well-known example is the effect of music. Vibrations from music have been shown to stimulate plant growth. Plants and animals are sensitive; they respond to emotions and environmental cues."

"You mean a positive environment impacts all living systems?" I asked.

Ankita jumped in, "Sir, let me share something I read on a website. It talked about how plants respond to human emotions. Dr Kim Johnson, a research fellow at the School of Biosciences at the University of Melbourne, studied plant senses and explained that one of the main ways plants communicate and respond to their environment is through chemical signals. Depending on what they need or the stimuli they face, they release specific chemicals into the air or soil. These chemicals can even influence the behaviour of nearby plants and animals."

She paused, then added, "Before that, in the 1960s, a polygraph expert named Cleve Backster connected a polygraph machine to a houseplant. He discovered that the plant reacted to his thoughts and emotions, suggesting it could sense them."

"Exactly," Amar said. "When we work with people or plants, we need to do it with happiness and purpose."

"Sir, I once read a quote of Aristotle: 'Pleasure in the job puts perfection in the work'," Ankita said.

"How do I make sense of all this?" I wondered aloud.

Viswa turned to me and said calmly, "That is the spiritual dimension of leadership. You can only generate positivity around you if you work in a positive culture. You cannot, and should not, go to work carrying anger, grief, stress or sorrow. Those emotions ripple through every interaction you have. A truly effective and inspiring leader learns to manage their own emotions. No wonder emotional intelligence is considered one of the key pillars of organisational success."

"You seem to be quite ambitious about your work culture," I observed.

"You used the right word, Bala," Viswa replied. "It's not just the work; it's the work culture that matters. Once you dedicate completely to your work *manasa, vaacha, karmana*—which means by mind, speech and action—it becomes a form of yoga *asana* (posture). No wonder our ancient scriptures say, *Yogaha karmasu kaushalam*, meaning 'Yoga is skill in action'."

Ankita laughed. "Bala sir, don't provoke him! If you do, he'll launch into a three-hour lecture on spirituality in work, and, trust me, it's super exciting!"

We walked down a muddy path, the three youngsters carrying some farm tools. I was delighted to see these young graduates, with their rural background, so grounded in their local culture. I mentioned this to Viswa.

He smiled and said, "These are the real youth who are honest about what they do. No hangovers, no hiccups. They face life head-on. They don't care about wearing jeans or keeping up with modern consumerist trends. They're self-aware."

"We don't care how others see us," Shyam added. "We're happy being ourselves."

"You know, Bala," Viswa continued, "most of our youth can be empowered. The problem is they lack role models. They don't have belief systems to guide them. And unless you help build belief systems that are relatable and relevant, they'll keep drifting."

"Sadly," Amar said, "the false belief systems created by marketing forces make it easy for people like us to fall prey. Many of my young friends have been emotionally hijacked, and they crumble under their own weight."

"Sir, we have a huge opportunity to nurture young leaders, especially from rural backgrounds. All we need to do is demystify their misgivings. We need to instil a belief that our country and our practices are in no way inferior to others," said Ankita, her eyes lit with passion.

"Leadership is contextual to culture. Isn't it?" asked Amar.

"I don't know," I replied.

Viswa laughed. "The values of leadership may be universal, but the practices are definitely contextual. There's a lot we can learn from the wisdom of other countries too."

I murmured softly, *"Aa no bhadrah kratavo yantu vishwatah."*

"Hey, what did you say?" Viswa asked excitedly.

I smiled. "Just quoting the Rig Veda: 'Let noble thoughts come to us from every corner of the world.'"

"Absolutely!" Ankita chimed in. "The thoughts we're surrounded by greatly influence us. They shape the vibes we breathe. I once read somewhere that 'Farming is a profession of hope'. It keeps our desires and dreams alive. It teaches optimism."

"We've also learned to accept mistakes and failure," said Shyam. "In farming, so much is beyond our control. Even when we take all the right precautions, nature often teaches us unexpected, and sometimes painful, lessons."

"Crisis management is an essential leadership skill in farming," Viswa said with a smile. "It's something we must explore more deeply. Farmers often rely on intrinsic, intuitive skills to manage crises, skills that are worth reflecting on."

As usual, Ankita had a quote ready. "You know, Abraham Lincoln once said, 'I am a firm believer in the people. If given the truth, they can be depended upon to meet any national crisis. The great point is to bring them the real facts'."

I was taken aback by how this serious discussion on leadership, often seen in management classrooms, was being deliberated with such ease and clarity. And that too, on a farmland!

Just then, a voice interrupted.

"Sir…" A middle-aged man approached and bowed to Viswa. "*Vanakkam* (greetings), Sir."

"*Vanakkam*," Viswa responded warmly.

The man, in his mid-40s, was wearing a *baniyan* (vest) and a *lungi*.

"Oh, Muni, how are you? It's been a long time since I saw you," Ankita greeted him with warmth.

Muni flashed a captivating smile and greeted the three youngsters. They stepped forward and touched his feet. I was surprised by the kind of respect they showed him.

"That's Muni," Viswa said, turning to me. "He's the undeclared vice chancellor of our learning university. He's a walking encyclopaedia on farming. He can talk to every plant, and every plant seems to respond. They smile and sway when he walks by. He speaks to them with love. To them, he's the leader of leaders."

I had never heard such encomiums given to a person who had not received formal education.

Muni, a bit shy, said humbly in Tamil, "Sir, I'm an illiterate. I only went to school until Class Five. Nothing more."

"How does that matter?" responded Viswa. "He taught me the nursery rhymes of agriculture! Even students from the agricultural college come to him when they want experiential knowledge that textbooks can't offer."

"Leader of Leaders"—the phrase kept echoing in my mind. "Does a leader need to be formally educated in an institution?" I asked, intrigued.

Ankita remarked, "Sir, many people who graduate from branded institutions of learning lack a basic connection with Mother Nature. They are verbose, speaking in pompous language that doesn't touch either the mind or the heart."

Muni continued, "After leaving my school, I've spent nearly 35 years with this land. Mother Nature has been my teacher and has taught me so many hard lessons. But I've always found that she has a solution to every problem. All you need to do is to look for the answer. And use common sense."

I wondered. He went on, "Sir, for one problem, there may be many solutions. Not all solutions will fit the same context. One must choose the right solution for the right place or right occasion."

I was amazed. "One problem can have many solutions. Isn't that the base of lateral thinking?" I looked at Viswa, who smiled and nodded. "You're right. That's lateral thinking. See how he talks about the power of choice in simple language."

After a few more minutes of discussion with Muni and the case studies he shared, it struck me that he should have been a professor at a college. "Leader of Leaders"—isn't that what he was?

Amar spoke up, "Sir, I recently read a book titled *Leading by Example* by Richard A. Conlow, and Muni just fits into everything it describes. While his methods may be rough and unrefined, at the heart of everything he does is a genuine sense of purpose, simplicity and common sense."

"Amar is right," Ankita interrupted. "Good leadership doesn't always require great policies and evidence-based reports. Muni is a living example of experiential learning, evidence-based reporting and authentic narratives. We respect him just as much as we would an institutional head."

Muni sensed that something was being discussed about him in a language he did not understand, but he displayed no ego or self-pride.

On Viswa's advice, Muni opened a cupboard and pulled out a score of small bottles filled with different types of soil. "This is acidic soil, this is basic soil and this is clay. Here's red sand, and this is black soil," he said, showing the samples and explaining what kind of crops would grow in each. His words reflected years of experience and seemed to carry the same warmth as if he were introducing the members of his own family.

"As much as fertile soil is needed for good cultivation, a strong base is necessary for leaders to grow. That base is a blend of courage, confidence and conviction. The younger generation can grow into great leaders only when they understand their culture and what their goals are," said Viswa.

"That brings me to an interesting question," I mused. "Do we recognise people climbing the leadership ladder through their experience, or do we only value those with degrees from higher institutions of learning, even if they lack practical experience?"

"It is a tough question to answer. Both paths have merits and demerits," Viswa replied. When it comes to process lines, productivity, marketing, and sales, many persons with groundbreaking experience have shot to fame as leaders. Nevertheless, there are also young, bright minds who think differently and are willing to lead organisations with innovative strategies. An organisation must make the right choice depending on its specific needs."

I found myself wondering where Muni would fit in the leadership spectrum. What category of leadership should I even place him in?

> **REFLECT**
>
> 1. Learning and leadership are interdependent.
> 2. Leadership is about creating joy in work and its environment.
> 3. Leaders need to have a great emotional intelligence.
> 4. Leaders need to keep their minds open to new ideas.
> 5. Leaders should learn to manage crisis with confidence.

YOUR LEADERSHIP QUESTIONS

1. How much time have you spared, or do you intend to spare, for your learning? What are the avenues through which you seek to learn?

2. How gracefully have you accepted your failures in the past? What initiatives did you take to get out of the fear of failure?

3. Do you have the entrepreneurial spirit? How would you like to showcase the entrepreneur in you?

4. "Pleasure in the job puts perfection in the work." Do you agree? How would you like to add pleasure and perfection to your work?

5. How would you like to create a positive working environment for the members of your team?

5
Farming: A Lesson in Inclusive Leadership

*The greatest leaders know that success is not about them,
but about the people they lead.*
—**Simon Sinek**
Author and speaker on business leadership

"Take a right, Bala. Our farmland is just over there," Viswa pointed, showing me the way. We walked along a narrow muddy pathway between two fields, each with different crops growing.

The three young people were walking ahead with so much enthusiasm, as if they were on a spacewalk. "Interestingly, the plants that grow in this part of the soil don't grow there. Even though both pieces of land are quite close, the types of crops you can grow here vary," Viswa explained.

"Just like two children born to the same parents are different in their aptitude and attitude," said Amar.

"And the soil provides the same nutrients to both of them, without any difference," added Ankita.

"Or do we say it's like two people with the same qualification but displaying different aptitudes and talents?" I reflected.

"There's often a lot of misunderstanding between the terms 'aptitude' and 'attitude'. Which of them takes you to greater altitudes?" Shyam asked.

"Insightful question," I murmured.

Viswa responded. "You need the right combination. If you have the right aptitude but lack the right attitude, you'll have a problem. Similarly, having a right attitude but no aptitude is pointless."

"Wow, what an observation!" exclaimed Ankita. "It reminds me of a famous saying by Zig Ziglar, the American author: 'Your attitude, not your aptitude, will determine your altitude'."

"I think that defines what leadership should focus on," said Amar.

"Amar and Ankita, this may be a very sweeping statement. It could be said differently. Every type of soil has nutrients that provide the necessary resources for the kind of seed it can host," said Viswa.

"Thank you, sir, that's a great learning for me," Ankita said.

"It is like the psychological profile of every child is different, and they need to be nurtured accordingly," Amar observed.

"And it's also like saying the leadership profile required for one particular field may differ from another, and therefore, you need to nurture them differently," added Ankita.

Muni, who was accompanying us and following our discussions closely, suddenly remarked, "This crop needs the support of urea, and the other side needs NPK." (NPK stand for nitrogen, phosphorus, and potassium.)

All of us laughed together.

I asked Ankita, "Don't you think Viswa essentially said the same thing you did?"

"No, sir. He showed me a better way to communicate. He really helped me see that there's a different perspective to what I said," she replied. I was pleased with the way the discussion was unfolding.

"And leaders need to have this multisensory reception mechanism. Am I right?" Ankita said, laughing loudly. "They should have eyes and ears everywhere."

"Celebrating perceptions is indeed an effective approach to leadership, as sir used to say," Shyam chimed in.

"Why do you say so?" I asked Shyam.

"Sir, receiving and celebrating the perceptions of others gives them an identity. It tells them that we respect their views, which enhances their self-respect. They feel quite at home, and this helps them gain an empowered confidence, with the feeling that they do matter. Leadership, in this way, becomes more inclusive," Shyam answered.

"You mean to say that this is an exhibition of inclusive leadership?" I asked him.

"Inclusive leadership?" shouted Shyam, "Just let me search this out." A few seconds later, he shouted again, "Ah! It's there."

Shyam then read out a beautiful quote from Nellie Borrero, the managing director and senior strategic advisor at a corporate company, "Diversity is a fact, but inclusion is a choice we make every day. As leaders, we have to put out the message that we embrace and not just tolerate diversity."

I responded, "Further, it's about giving people the freedom to be who they are and what they are. This helps them to step out of their shells. Recognising and appreciating team members help them grow and boosts their positive energy."

"Sir, what are the basic traits of an inclusive leader?" Shyam asked.

Viswa responded, "Listen. Inclusive leaders generally exhibit a lot of humility. They are more humane than others, showing a greater willingness to accommodate. They have a high level of cultural intelligence and are ready to make personal sacrifices to support others in their life and work."

"Do you think the freedom these leaders give and the tolerance they show will be taken advantage of by their team? Will it affect the organisation's discipline?" I asked.

"No sir. Freedom comes with accountability. It's not about being arbitrary. The moment someone realises that their words carry weight, they'll be careful to provide credible inputs. They'll take responsibility for their words," Shyam replied.

"Could this somehow encourage stubborn or aggressive behaviour?" I intervened.

"Sir, you're being a bit judgemental," Ankita reacted.

Amar then said, "Possibly, it's the bias we bring into the relationship. Often, we not only harm relationships because of our selective bias, but we also tend to act superior to others. Am I right, sir?"

Viswa was listening patiently and then said, "Do you see their intellectual levels in the statements and arguments they advance?" I just smiled.

"We learnt it from him. He used to repeat John C. Maxwell's quote, 'A leader is one who knows the way, shows the way, and goes the way.' Our minds have been enriched by his thoughts…"

"Like the soil that is enriched by Mother Nature," Shyam intervened.

"But to a common man, the soil might look more like dirt," I said.

Everyone laughed. All eyes turned to Ankita, eager to hear what she would say.

"That comes from a prejudiced eye. It's the perspective of someone who thinks they walk on the soil with their feet. We believe we walk on it with love and affection. You know, sir," she continued, "Oliver Magny, who runs a wine business, wrote in his book *Into Wine: An Invitation to Pleasure*, 'Studying wine taught me that there was a very big difference between soil and dirt: dirt is to soil what zombies are to humans. Soil is full of life, while dirt is devoid of it.'"

I stood there, dumbfounded. I realised that leadership is like farming—an experiment, a constant learning process. The lessons in this field are endless. I learnt that both the breeze and the storm offer valuable teachings. I learnt that both the soil and the dirt have their lessons to impart. I learnt both the rain and the sunshine hold their own wisdom. I learnt that both drought and flood provide insights. I learnt that both the seed and the harvest teach important lessons.

"Bala, you seem to be lost in your thoughts," Viswa said, gently touching my arm.

"Unfortunately, even in our social and professional setups, some leaders look at their followers and fellow beings as dirt," I replied.

"Social inclusion and social justice, therefore, are crucial in defining and developing leadership. A leader who is blind to these or lacks sensitivity will eventually be rejected by the system," Viswa stated.

"But sir, in trying to address some of these issues, are we compromising on quality?" Ankita asked.

"The responsibility of a good leader is to nurture and develop quality, not to marginalise the workforce or followers as substandard," Viswa responded.

"Quality comes with learning and experience. Therefore, a good leader creates opportunities for the team to learn and experience," he added.

Muni, who had been listening to these discussions sitting quietly in the corner, smiled. "My farmers learn only through experience," he said in his native language.

I responded, "But how much time, money and resources do we invest in them? Aren't we losing out on costs?"

"Developing leadership does have a cost, but it's an investment. A good leader should include this as part of their executive programme," Viswa answered.

Shyam then intervened, "What about learning for the new recruits?"

Viswa responded, "Every member in the organisation, including the leaders, is on a learning curve. They need to be nurtured and mentored periodically to remain fit. Professional fitness is as important as physical fitness."

Muni clapped, as if he were listening to a political speech. Everyone laughed.

"Honestly, Viswa, I thought I was coming for a holiday to spend time with you. I never thought I'd be coming to a university to learn. The farm is indeed a university, teaching you every second, at every step," I said.

"To put it differently, Bala, it's a university that is willing to teach those who are willing to learn. And there are five vice chancellors in this university," Viswa explained.

"Five?" Ankita asked.

"Yes," Viswa replied. "They are the five elements—fire, water, earth, space and air."

Amar joked, "And what subjects do they teach?"

Shyam responded, "Agriculture and Life—both these subjects have all the five elements as integral to their basic concepts."

Viswa intervened, "Any discipline—from A to Z. If you have the eyes to read, ears to listen and the hands to work, the farm is one field that can teach you everything."

"Sir, is it also a management institute?" Ankita smiled.

"Don't ever limit the universe of agriculture to an institute," Viswa replied. "It's nature's learning university. Not only management and leadership, but also history, geography, ecology, war and peace. So many wars have been fought for this land and many more are fought on it."

"Wars?" asked Amar.

"Yes. Even now, communities and states fight over land possessions. It is nothing short of war," Viswa said.

"True, sir. Even states fight over the quantum of water that flows from one place to another," Amar said.

"Do you mean to say a good leader should always be ready for a war?" I asked.

Viswa laughed, "You know, Bala, there's a saying by Leon Trotsky, the Soviet revolutionary and political theorist: 'You may not be interested in war, but war is interested in you.'"

"How true!" Ankita exclaimed.

"And a good leader sees opportunity even in chaos? Am I right?" I asked. Everyone nodded.

"I think we need to think a little more progressively," Viswa continued. "We cannot talk of land in exclusion. All the elements coexist—water, air, space and…" He didn't finish, but Amar responded, "Holistic?"

"Yes," said Viswa. "Mother Earth is one. You cannot treat different parts of her beautiful soul in exclusion. One of the problems in our thinking is that we often approach problem-solving as if it's a solution to an end goal or an achievement towards an aspiration. We don't see problems as part of an existential quest for peaceful and interdependent living."

"Haven't we reduced our thinking about the land to just a material that gives tonnes of produce, thus focusing solely on its economics?" I asked.

"Agreed. We need to broaden our view of farmlands as a spirit that coexists with us. It's much more than just statistics," Viswa responded.

Amar then drew everyone's attention to an important research paper. He said, "In one of the research papers on phototropism and gravitropism, I read the following observation:

> Light and gravity are two of the most important environmental parameters affecting plant growth and development. To maximize available light and nutrients, plants orient their stems towards the direction of illumination and away from the gravity vector, and, conversely, orient their roots away from light and towards gravity. In plant organs, there may be a competition between gravity (gravitropism) and light (phototropism). At the same time, the truistic signalling and/or

responses induced by both stimuli may be independent processes that coexist, or they may modulate each other.

Amar added, "That's indeed similar to the growth profile of a human being, who grows both inwards and outwards. His outward glory and profile depend on his inner strength, much like plants."

"Wow!" I exclaimed. Everyone clapped at his observation.

"True indeed!" said Viswa.

"That seems to explain the need for a growth mindset," I remarked.

"A leader is someone who has a growth mindset," said Amar.

"There are no options," said Viswa. "Learning, unlearning and relearning are critical to a growth mindset. And farming provides that growth mindset through experiential inputs."

"But don't we look at farmers as people with a stagnant mindset, rather than one with a growth mindset?" I asked.

"I disagree," said Viswa. "Their understanding and definition of growth revolves around sustainability. They don't force growth; they understand it. They witness it and experience it. The process of growth doesn't cause them stress. There's a wonderful saying by Lao Tzu, the Chinese philosopher and writer: 'Nature does not hurry, yet everything is accomplished.' Farmers truly understand the spirit of this statement."

"You are right. Unlike corporates, farmers view growth not just in terms of numbers or productivity, but as an integral part of the well-being of a process. Often, they consider growth in the context of sustainability," Shyam responded.

"Well, getting back to the growth mindset, what are its characteristics?" I asked.

"Amar will lead this discussion," Ankita teased her friend.

"A growth mindset is a positive mindset, one that is willing to identify, seek and engage with opportunities. It is transformational in nature, so it's dynamic," said Amar. "A growth mindset is reflective, minimising waste, loss and non-productive time and resources for meaningful recycle."

"Transformational and dynamic. Excellent," I acknowledged.

"But Amar, tell me. Could it be dynamic and transformational, or is it just one of them? Because if something is not dynamic, it need not be transformational. I'm trying to understand the logic," said Ankita.

"You are clever and shrewd," Shyam complimented Ankita.

"It feels like I'm sitting at a brainstorming table," I commented.

Viswa laughed. "Brainstorming… a big word indeed. Unfortunately, many aspiring leaders don't know the difference between conferencing, discussing, brainstorming and conversing. Folks, you need to know the difference. If you want to emerge as a good leader, you need to know the difference. Not all talks are brainstorming."

"Brainstorming indeed causes a creative tension among the members," I explained.

Ankita jumped from excitement. "What? Creative tension? I'm hearing this term for the first time. Can you explain it?"

"Yes," I replied. "It's a discussion that leads to creative tension, where there's a joint, collaborative, focused and purposeful debate on different perspectives of the problem, its solutions and its future impacts. It calls for higher-order thinking." I paused for a moment, before adding, "You know, Peter Senge, the management guru, says 'Mastery of creative tension brings out the capacity for perseverance and patience. Time is an ally.'"

"Sure, sir. We'll come back after doing some research," said Amar. "I believe we need to understand the meaning and use of creative tension as a positive force for creativity and growth."

"Creative tension may also be experienced during problem-solving and crisis management," Ankita observed.

"That's it. A research-oriented mind often experiences creative tension. Unless you've developed such a mind, you won't be dealing with authentic information, facts and figures," I tried to intervene.

"Is a research-oriented mind the same as the practice of research?" questioned Shyam.

"A good question. A research-oriented mind is indeed an active, inquisitive mind—one that seeks data, looking for verification, is willing to investigate and is eager to expand the scope of current knowledge. There are well-laid-out steps for both the process and methods of research. A researching mind should follow those basic guidelines," I explained.

"Viswa sir, can farmers also acquire a research-oriented mind?" asked Shyam.

"Why not? Any human mind can be a research-oriented mind, provided it has the right methodology," answered Viswa. He continued, "Albert Szent-Györgyi, a Hungarian biochemist and Nobel Prize winner said that 'Research is to see what everybody else has seen, and think what nobody else has thought.' Many farmers have a keen sense of observation, alongside intuition. They do have a researching mind," Viswa responded.

"Do you think that a research-oriented mind can be developed from a young age?" I asked.

"Yes. Neurocognitive scientists believe that," Viswa replied. "Training the mind to become an inquiring mind—a mind that is curious, willing to question and able to perceive things differently."

"Well, we have a long way to go. But for now, the farms have given us plenty of opportunities to learn a lot about leadership," he concluded.

> **REFLECT**
>
> 1. An effective leader should possess the right attitude and a positive aptitude.
> 2. A growth mindset is critical for developing leadership.
> 3. Inclusive leadership is a humane approach that seeks to involve people in learning and growth.
> 4. A leader with a research-oriented mind becomes future-ready.
> 5. Leaders must understand that in the midst of chaos, there is also opportunity.

YOUR LEADERSHIP QUESTIONS

1. If you were appointed as the team leader of a project, what steps would you take to develop the right attitude among your team members?

2. What gaps do you find in your leadership profile that may hinder inclusive leadership? What strategies would you adopt to bridge these gaps?

3. Do you have a growth mindset? If asked to provide evidence of it in an interview, how would you respond?

4. Assuming you are going to have a brainstorming session with your team on crisis management issues, what preparatory notes would you prepare to initiate and lead the session?

5. As a team leader, how would you ensure there is no "blame game" or "passing the buck" among team members, and that the focus remains on problem-solving and crisis management?

6
The Soil is the Universal Mother

The nation that destroys its soil destroys itself.
—*Sir Franklin Roosevelt*
Former President (32nd) of the United States

As we walked with our group, I noticed a small house. It had a single bedroom, a sitting room and a pantry. Outside the house, there was a shed with a cupboard to store tools and other items. Apart from the regular cot and mattress in the bedroom, there were two cots made of coconut fibre placed outside.

"This's a small resting place. I come here sometimes to stay overnight," Viswa explained.

"We've also kept him company," Amar and Shyam added.

"Blessed they are," Ankita smiled.

"There're a few mud pots here, just like ones you see in my house. Each one has clean, potable water with some herbs," Viswa continued.

"Herbs?" I raised an eyebrow.

"Yes Bala. One's tulsi (*Ocimum tenuiflorum*), another has *ajwain* (*Trachyspermum ammi*), the third has *khus* (*Chrysopogon zizanioides*) and the fourth has an Indian herb called *Nannari* (*Sarsaparilla*). They're all good for health," Viswa added.

"Well, I know we used some of them as home remedies. My grandma would boil them in hot water and have us drink it," I said.

"Yes, these're also part of our heritage food and drinks," Viswa nodded. "People in the village generally avoid soft drinks, which are mildly acidic, or caffeinated beverages. The healthiest drink you can get here is fresh tender coconut water—straight from the tree. The most popular drink, though, is buttermilk," he added.

"And palmyra fruits (*Borassus flabellifer*) too. I just love them," said Ankita.

"Now, coconut water is processed and exported to other countries," I remarked.

"There's also a huge market for buttermilk," Viswa added.

"True, we undervalue them and end up falling victim to artificial and processed drinks," I responded.

Viswa smiled and said, "We've a separate farm in the village where we grow herbs. We've farmers who are experts in taking care of them. There's a growing demand for these herbs, and the farmers are able to make decent profits from cultivating them."

"Increasingly, many of these herbs are becoming the basic resources for several allopathic medicines. The pharma industries need support from these farmers," Ankit commented.

"I think our corporate leaders should have the power to change the course of thinking to support and facilitate native natural products," said Amar.

"Sir, many of our corporate systems are imprints of Western models. They've been built on the footprints of Western culture, and so they believe that wearing suits and eating packaged food defines modernity," said Shyam.

"I may not agree fully. Yes, they follow the footprints of Western culture, but wearing suits and eating packaged food aren't necessarily signs of changing times. This phenomenon is global," I argued. "Let's not link it to other native issues."

"To say that every aspect of modernity is tied to Western culture is taking things too far. That approach is wrong. For centuries, Bharat has celebrated inclusivity. We need to understand that we live in an inclusive and interdependent world," I continued. "The growing food fad is one of the outcomes of business dynamics. Children and youngsters today are moving away from traditional food and gravitating towards packaged, pre-processed foods. The reason is the myth created around them."

"Why so?" asked Ankita. "Is it because we haven't thought about re-imagining the Indian food system? Or is there something more that needs to be done to refresh the system?"

"Listen, in most business and cultural platforms, the dominant force calls the shots. They project that those who do not consume their products or services are outdated and sub-standard. Creating emotional need and loyalty seems to be the two dimensions of the modern business game. From flying to banking, loyalty programmes are everywhere, drawing in consumers. The consumer often loses track of the quality or cost of a product, just to stay in the game and rack up loyalty points for insignificant rewards. So, the rules of the game appear to be different," I responded.

"I think the winds are changing," I added. "There's a lot of advocacy around traditional food these days."

"One has to find the game changers," Viswa observed.

"Are leaders essentially game changers?" asked Amar.

Viswa smiled. "They don't always have to be. Game changers are required only when there's a real need and appropriateness. Engaging with game-changing processes where they aren't required or making frequent game-changing interventions can shake the basic edifice of any system or a process. It takes intelligent application."

"It's important that we ensure the quality of our products and focus on packaging that keeps them safe, healthy and increases their longevity. Even in our country, many food processing and packaging industries have come up, which are globally validated," I said.

"Yes, that's true. But we still have a long way to go," said Viswa. "One of the major challenges our farmers face is the lack of access to direct markets. The role of middlemen really robs them of their legitimate profits. In spite of several governmental interventions, the problem persists."

"That really means we need to change the way we think. We need more influential advocacy that can impact and synergise both thought and action. We should encourage more farmer markets, like they have in other countries," he added.

"Well, is leading by thought and leading by action more important than other forms of leaderships?" I wanted to clarify.

Viswa laughed. "I'll tell you an interesting story. When I was in college, I represented the college in debates. I participated in several inter-collegiate competitions. In one such competition, I was given an extremely interesting topic to debate. The topic was given to us only 30 minutes before the event started, and it was: 'Ideas govern the world, not men'. We could argue either side."

"Indeed, that's interesting," I said. The three youngsters gathered around.

"Gandhi or Gandhism? Marx or Marxism? Seems like a fascinating discussion," said Shyam.

"It's like the proverbial 'Chick or the egg?' question," interjected Ankita.

"True. Either way, you appear right. It all depends on how you build the argument," Viswa continued. "A man lives by the thoughts he generates, and then through those thoughts, he lives for generations by performing the relevant actions."

"Thought leadership is definitely a critical and non-conservative way of looking at leadership," I suggested.

"Is it revolutionary?" asked Amar.

"Yes, in a way," I replied. "A thought leader not only thinks differently but also ushers in new ideas for consideration and practice".

"Thoughts are generated and lay dormant in the mind, for different periods, depending on the person. And then, there are the various stages of thought dynamics—thought articulation, thought cultivation, thought processing, thought management, thought navigation, thought re-engineering and thought cultures. The thought matrix comes to life with its three-dimensional parameters. There's so much to reflect on when considering the vector of thought dynamics to ensure a right type of thought leadership," Viswa said.

"Do our mid-level leaders have enough time or opportunities to become thought leaders?" asked Amar.

"You just can't make or train a person to be a thought leader. It comes from within. Thoughts blossom naturally. Thought leaders are generative by nature. They don't get easily influenced.

They're passionate about their thoughts, emotions and feelings. Unfortunately, these leaders often fall victim to the rat race—chasing numbers, chasing achievements. If they don't fit within the operating universe, they become isolated. Those who survive these onslaughts eventually become role models, influencers and icons," Viswa said.

"Are thought leaders generally disruptive?" asked Shyam.

"Maybe, or maybe not. They could be visionaries. They could be critical thinkers. All of these skills can broadly fall under thought leadership," Viswa said.

Ankita, as usual, caught everyone's attention with a quote: "According to author Pearl Ziu, 'Thought leadership is by nature evolutionary, in that it must always be part of an ever-evolving flow'."

Viswa let out a long sigh. "What happened?" I asked, noticing his expression.

"When I first came to this village, even my simplest ideas and suggestions were seen as disruptive. I had to put in a lot of effort to convince the farmers," Viswa said, looking at Amar.

Amar quickly asked, "Sir, what are the features of disruptive leadership?"

"Disruption is a break, an intervention, a disorder, a deviation from a routine continuous system or process, a thought or a dynamic. It brings about change, transformation, innovation, re-engineering or restructuring of the existing state. It changes the dynamics of the system—thought, style, approach, objective or target," Viswa responded.

"Is disruption good?" Amar asked.

"Whether something is good or bad depends on how you perceive it. Disruption often acts as the catalyst for change.

Take the Green Revolution, for example—it was a direct result of disruption. Disruption is essential for transformation and progress," I remarked.

"What about the White Revolution—the cooperative movement in milk production and supply? That, too, was a disruption that introduced the concept of cooperative production and marketing," Ankita observed.

"You're right, sir. People generally don't want to change quickly. Everyone prefers comfort zones. They don't realise that times are changing, and so should our thoughts," Shyam reacted.

"One of the most successful disruptions in recent times is the rise of the 'digital economy'," I said.

"Many call these discussions academic and non-productive," said Shyam.

"Every organisation should have a team of 'Idea Managers'," suggested Amar. "Their exclusive role should be to generate ideas and explore possibilities," said Amar.

"Interesting. But is it feasible?" asked Ankita.

"That's the real starting point for research and development (R&D) in every organisation," Amar responded.

Viswa was listening quietly, wanting the team to engage in these discussions without any hesitation.

"It feels like we're going overboard," I said with a blink.

"The entire process of farming requires a keen sense of observation, data collection, analysis and strategic interventions that are relevant, contextual and comprehensive. In short, several of the processes we use in research are fundamental to farming," said Viswa.

"It's fascinating to see farming evolve beyond the traditional model. I'm not sure if society has deliberately branded farmers as intellectually inferior, but the superiority that other professionals show towards farming is both incorrect and in need of social correction. With modern technologies supporting farming, we also need to work towards social upliftment of our farmers," I commented.

"Farmers and teachers need to be celebrated. The former brings food for the stomach, and the latter brings food for thought," said Ankita.

We all nodded in agreement.

> **REFLECT**
>
> 1. Leading by action and example takes a leader on a forward-moving journey.
> 2. Thought leadership helps create and celebrate diverse perspectives.
> 3. Informal learning is an equally valid gateway to leadership, just like formal learning.
> 4. Understanding context is crucial for effective leadership.
> 5. Disruptive leadership helps usher in a fresh lease of life for processes and systems.

YOUR LEADERSHIP QUESTIONS

1. If you were to lead an organisation, what kinds of initiatives would you take to provide continuous experiential learning for your colleagues?

2. The role of an idea manager is important for any organisation. As a leader, what broad guidelines would you give to this person?

3. To strike a critical balance between innovation and resistance to change, how would you plan for the future of your organisation?

4. It is said that a disoriented leader can lead an institution to disaster. What common mistakes do leaders make that cause disorientation in organisational dynamics?

5. What, in your opinion, defines the profile of a 'Leader of Leaders'? List three major challenges such a leader typically faces.

7
The Power of Seeds: Ownership Values

*If you can look into the seeds of time,
and say which grain will grow
and which will not, speak then unto me.*
—William Shakespeare
English playwright and poet

"It's time for some refreshment. Would you like to have coffee or tea?" Ankita asked.

"Well, I'll bring some tender coconuts for all of you. That'll give you the energy," said Muni.

"I'm quite impressed with the way you celebrate the soil. It's truly a form of reverence for Mother Nature," I continued the discussion.

Ankita chimed in, "Sir, do you know what William Shakespeare, the great poet, said? He said, 'One touch of nature makes the whole world kin'."

"True. It's important to celebrate and respect the soil. If the soil isn't healthy or properly cared for, growth will definitely be affected. Similarly, it's important to take care of the intellectual and professional landscape of any organisation to ensure its growth can be taken care of," Viswa replied.

"Can you explain that in a bit more detail?" I asked.

"Bala, as I said, every organisation needs the right intellectual and professional environment for its leaders to grow," Viswa explained. "An organisation, no matter what it does, shouldn't just be viewed as a vehicle for business, manufacturing or trade. Its real strength lies in how it nurtures human resources. That's what enables long-term growth and success

"Many people join organisations not just for high salaries but because they offer opportunities for learning, development and personal growth. They see this as a way to enhance their self-worth. That's how people begin to celebrate themselves. The feeling of 'I can' and 'I matter' gives them the real sense of purpose. In such organisations, employees tend to show stronger loyalty. Their engagement improves in quality, and the pursuit of excellence becomes a shared agenda."

"You're right. People used to work in the same organisation for many years just because they loved the organisation, whether or not they got any recognition for their long service," I said.

Ankita smiled and said, "My grandfather used to wear a watch he was gifted when he completed twenty years of service in the same organisation."

All of us had a hearty laugh.

"Times are changing," said Amar.

"Of course, the comfort level while working in an organisation is important," said Shyam.

"Farmers have a similar feeling when they work in the field. With every single step they take, they feel their comfort level. They feel the soil, they feel the plant, they smell the air, they enjoy the dry land and the moisture—and they make their decisions. They get the feeling 'everything is right' or 'something is wrong' with every step they put forward. It's like living with the system," Muni spoke in Tamil.

I wondered how Muni sensed the direction of the discussion.

"In many organisations, leaders show personal care and attention to the members of the team. They know their teammates' family profile well," I said.

"But Sir," Amar intervened, "With more professionalism coming through, personal touches have largely disappeared. We've started looking at humans like machines that just function."

Shyam remarked, "It's just like the difference between tilling soil manually and using a tractor. The feel is different," Viswa smiled at his sensitivity.

"But a good leader bridges the gap between management and employees by playing a positive interlinking role," I opined.

"There's an increasing focus on HR departments in organisations, and they're expected to take care of the welfare and growth of their employees," Shyam added.

"It's indeed a two-way process: both the employees remaining committed to an organisation and the organisation taking care of the employees and their families," Viswa said.

"I think there are already enough rules by the government regarding this," Ankita remarked.

"Rules alone don't help. An organisation needs to develop its own culture of employee care. And this positive attitude should percolate from top leadership to all other levels of leadership like a cascading process. The organisational culture impacts how employees feel and engage with the organisation," Viswa said.

"A leader doesn't stay put in one place. His leadership strategies walk through the entire organogram," I felt.

"What about Townhalls in corporate companies?" I asked Viswa.

He smiled and said, "In many corporates, it's a periodic ceremony conducted professionally. But in many such events, the honesty of purpose is lacking. It becomes more of a social gathering, a way to officially communicate a message that 'we are listening'."

"Do you mean to say there's no ownership on the part of the leadership in such activities?" Amar asked.

"You can't say that, but it becomes more procedural—a way to mark off the event that's been conducted ceremoniously. It's just on record to justify the role of leadership in any future litigation," I said.

"Interestingly, in villages, all the people working on a farm are considered family members. Landowners even participate in the family functions of their workers. The leaders in the farmlands show a kind of ownership over the welfare of their co-workers," Viswa said.

"Cultural camaraderie is seen more in villages. Though there may be differences, people tend to keep their boundaries to avoid aggressive conflicts," he added.

Just then, Muni walked in with a couple of glasses of tender coconut water and some white bits of tender coconut pieces. They were indeed inviting.

"Thank you, Muni," all the four said simultaneously. I didn't.

Visha looked at me and said, "We always make it a point to say 'thank you' to anyone in this village when they do something for us. You should see their face when we say that. They're so delighted. I think that's how you win people over—a lesson in Relationship Management."

Muni smiled and said in Tamil, "The entire village feels so happy when they hear the words 'thank you'. We learnt this from Viswa Sir."

"The celebration of even the last person in a hierarchy in the community elevates their outlook and enhances their sense of belonging. Unfortunately, many organisations have developed a hierarchical, authoritarian model where the concept of 'servitude' affects people at all levels," said Viswa.

"What other factors do you think contribute to developing leadership?" I asked.

Before I could finish the question, Muni brought a few bottles containing some grains and handed them to Viswa.

"This's what's important: the seeds…", Viswa began.

"Seeds?" I interrupted him.

"As the proverb goes, 'As you sow, so shall you reap', Viswa said, looking at me. "Seeding the right ideas among all your employees is critical to growth."

"Sir, a good seed may not grow well if there's a mismatch between the seed and the soil. Not all plants grow in all types of soil," Viswa said.

"Selection of seeds is both an intelligent and critical process for successful farming. Similarly, selecting the right type of people is important if you want them to grow in your organisation. If you don't do that, you'll be wasting resources, processes and productivity," he added.

"I agree. It's important to have people who have an aptitude for the work they do. Sometimes, people work for their salaries, not because they enjoy their work," I responded.

"Sometimes, workers are employed in an organisation on the recommendation of senior people, both within and outside the organisation. These people may not have the interest or aptitude. Over time, they become a liability to the organisation," I added.

Ankita smiled and said, "Sir, it's like putting any kind of seed in the soil just because you have them, without knowing whether they'll grow. To be more specific, there are different types of seeds—heirloom seeds, open-pollinated seeds, hybrid seeds, organic seeds and genetically modified seeds. Each type has different vitality, and each is contextual to geography, soil, location and purpose."

Viswa nodded thoughtfully. "See these four seeds: Amar, Shyam, Ankita and Muni. They have common purposes but diverse interests. They're passionate about what they do and willing to learn. I would say they're self-learners, and one step further, they're self-directed learners. You don't have to tell them what to look forward to. In fact, I learn a lot from them. For example, Muni was my teacher when I came here with very little knowledge of farming. He is my *Guru*."

Muni understood and smiled. "Don't use big words, Sir," he said in Tamil, gesturing with his hand towards the sky, as if attributing it all to divine will.

"I had read a story about seeds, and I'd love to share it with you all. May I?" asked Ankita.

"Of course, go ahead. Storytelling is the best way to motivate people," Viswa encouraged.

Ankita began, "A farmer was walking through his farm, throwing seeds all around. Two seeds fell close to each other. After some time, one of the seeds opened its shell and peeped out. The second seed, watching this, shouted, 'Hey, don't do that.

Stay safe within the shell. If you open up, you'll have to face the weather. It could rain, or it might be hot. You'll suffer either way.' But the enterprising seed didn't listen. In a few days, the seed became a tendril, stretching towards the sun. The dormant seed laughed and said, 'I told you! You're drying out. That's why you should listen to your peers.' As days passed on, the tendril grew into a plant. The rough weather and storm forced the plant to face hardships. Again, the dormant seed had a hearty laugh, 'See? How comfortable I'm in my shell while you're struggling to grow.'

"Days went on, and the plant started bearing fruits. It was weighed down by the heavy load of fruits on its head. The dormant seed peeped out again and said, 'I'm alarmed at your foolishness. Look at how much weight you're carrying! Isn't it an unnecessary burden?' Right then, a hen passing by saw the seed and picked it up. The plant, weighed down by its fruits, felt sorry for the seed that refused to live and grow."

"That's a story with a powerful message," I said.

"Hey, guys! What's the message you get from this story?" asked Viswa.

"Life is all about challenges. To grow, one must learn to live with challenges. Nothing should defeat one's pursuit to grow, which demands constant change," said Amar.

"Learning is all about opening one's mind to change. If one is unwilling to change, they will remain static. And any process that is static is a death sentence," said Shyam.

"Good interpretations," I said, impressed.

"Anything more?" Viswa asked. He then continued, "Listen. The first seed represents a leader who is conditioned by a survival syndrome. The second one represents a leader who is an

The Power of Seeds: Ownership Values / 75

entrepreneur. In the first case, the leader suffers from 'learned helplessness' due to past experiences, fear and an unwillingness to learn or prove one's worth. In the second case, constant struggle teaches the leader to learn and adapt from every experience. Further, the second seed represents a leader who isn't just changing but transforming every day based on learning. People look forward to seeing what he will become next."

Muni, who was listening attentively, stepped aside and returned with two small paper plates, each containing a sample of seeds.

"Thank you, Muni," said Viswa. "See how intelligent this guy is. He's brought two samples to illustrate the point. This one is a roasted seed. It won't grow. The other one is a fertile seed, so it will grow well."

"What does the roasted seed represent?" I asked.

Viswa laughed uncontrollably. The others looked at him curiously.

"The roasted seed is like people who are burnt out. They appear exhausted and depleted of energy. They look like they've lost hope and are tired of doing anything further. They resist change and have no aspirations to face challenges," Viswa explained.

Amar, Shyam and Ankita exchanged smiles, as if acknowledging Viswa's words. Muni nodded, as if he understood everything, and murmured, "Love, only love…" The words expressed his deep passion for the soil. Everyone smiled.

"Sir, I read somewhere that 'Farming isn't a battle against nature, but a partnership with it. It's about respecting the basic interdependence of nature and ensuring that it continues'," Ankita said.

"And that's true for every organisation and every system," Viswa remarked. "If you want to lead an organisation or a system, you don't battle against it or its constituents. You respect its nature and become an integral part of it, bringing about change and transformation. If you want transformation, you have to work with the system rather than 'on' the system."

As we spoke, we could hear the voice of the cattle outside the house. We moved out.

REFLECT

1. A leader ensures the intellectual and emotional well-being of team members.

2. An effective leader nurtures and mentors the organisation's human resources by promoting their sense of self-worth.

3. A leader empowers team members to weather challenges with courage, confidence, and conviction—and to grow into future leaders.

4. The "burnout syndrome" creates resistance to growth in any organisation.

5. Leaders don't battle organisations or systems—they work within them to bring about transformation.

YOUR LEADERSHIP QUESTIONS

1. There is a growing belief that older employees in organisations are difficult to transform. As a leader, how would you support their transformation and growth?

2. "Moderating change is as important as ushering in change." Do you agree? How would you approach moderating change in the organisation you lead?

3. Identify any three indicators of individuals experiencing "burnout syndrome." How would you work to help change their behaviour and restore their well-being?

4. Gratitude for support and service fosters goodwill among all workers. What procedures could you introduce to ensure team members feel recognised and valued for their contributions?

5. "I have not worked in this organisation for the salary they pay; I have lived with the organisation," says an employee who is being let go due to a technology transfer. As the leader, how would you respond to this situation with empathy and fairness?

8
Farmers: The Barefoot Leaders

A man who wants to lead the orchestra
must turn his back on the crowd.
—**Max Lucado**
American author and pastor

"I think it's time to move to the field," said Amar.

"Yes, yes…" Muni nodded. We walked out.

"Don't you lock the house?" I asked Viswa. He smiled. "No. Nothing much is there to be locked. It's open. We work with a system of trust," he said.

"Trust is an important requisite for joyful coexistence. A leader must instil trust in people and systems," he continued.

Ankita asked, "Sir, have you read the book, *The Speed of Trust* by Stephen Covey?"

I nodded. "Yes. I read it and I liked it. The way the author developed the concepts".

Ankita continued, "In that book, the author says, 'The first job of a leader—at work or at home—is to inspire trust. It's to bring out the best in people by entrusting them with meaningful stewardships, and to create an environment in which high-trust interaction inspires creativity and possibility'."

"And do you know how he qualifies the term 'Trust'?" Viswa asked me. I stayed quiet.

He smiled and said, "Trust—the one thing that changes everything."

"We have examples of several leaders who gained massive followings because people trusted them: Mahatma Gandhi, Martin Luther King Jr., Jayaprakash Narayan, Swami Vivekananda and many others, both in politics and public service," he added.

"If leaders are not trustworthy, they are subject to a quick fall. They fall from great heights, and it's not just a quick fall—it's a free fall," he continued.

"And sometimes, the effect is so profound that they can't rise again," Shyam added.

We all acknowledged the power of this sentence and laughed.

"Both in politics and in corporate world, there are many examples of people who have suffered a free fall. Although some managed to maintain a life of luxury afterwards, the community viewed them with contempt," Viswa remarked.

"Like dry fallen leaves?" Shyam asked.

"Yes. Interestingly, leaders who thrive exclusively by their charisma are subject to a free fall. As long as their charisma finds a face and popular value, they are at the top. Once there's an assault to their charisma, they get eclipsed by others," Viswa replied.

In a few moments, we stepped into the field. "Bala, please leave your chappals here. You need to walk bare feet. Only then will you truly feel the gentleness and generosity of Mother Earth," Viswa said.

"Is it safe?" I asked.

Viswa looked at me. "Were we born safe? Have we lived safely for centuries? While I agree that footwear provides us safety, it's equally important to feel the earth on which we are born, every now and then."

He chuckled. "I recommend you walk barefoot for at least a few metres. I know it's both painful and pleasurable."

I commented, "I recall reading a book in the library some years ago. The title was *Barefoot Leaders: Simple Effective Leadership* by James Caroline."

"Wow! What's that, sir"? Ankita asked curiously.

"Barefoot leadership is a conceptual model. Barefoot leaders are ordinary people who achieve extraordinary results with surprisingly few resources. They don't need fancy shoes or power jackets to get things done. They lead by serving people at the ground level. They understand the art and the joy of going an extra mile," I said.

"And what does the book say?" she continued.

"The principles in this book are derived from human behaviour, so they transcend organisational structure, size and industry type. This book helps enhance the outcome or achievement levels of leaders," I replied.

"Sir, have you met any of those barefoot leaders?" Shyam asked.

Viswa laughed uncontrollably. "An eminent author, Elin Hilderbrand, who wrote a novel titled *Barefoot*, once made a remark I couldn't agree with more: 'Being a mother was the best of all human experiences, and also the most excruciating'."

"Indeed. I believe every mother is a true example of a barefoot leader," Ankita said thoughtfully. "That's why motherhood is so often equated with divinity."

"But there's no professional touch to it," said Amar.

"Be careful with your words, Amar," cautioned Shyam. "We don't need to analyse everything with a professional lens. Just like the farmers we're celebrating as role models for leadership development, we need to see motherhood as a powerful and successful model for emotional leadership."

"I read somewhere," Viswa said, "that motherhood is 'the exquisite inconvenience of being another person's everything'."

"True," I continued, "and that tie perfectly with what Simon Sinek says in his book *Leaders Eat Last: Why Some Teams Pull Together, and Others Don't*, 'The true price of leadership is the willingness to place the needs of others above your own. Great leaders truly care about those they are privileged to lead and understand that the true cost of the leadership privilege comes at the expense of self-interest'."

"And that's exactly what mothers do," argued Ankita.

"The first name that comes to my mind when talking about "barefoot leaders" in India is Vinoba Bhave, the leader of the Bhoodan Movement. He walked across the country to encourage the redistribution of land from landowners and zamindars to farmers, so the land could be cultivated and better used. There are others too who have worked for similar causes," Viswa remarked.

"I've heard Vinoba Bhave's name, but I need to study more about his remarkable contributions," said Amar.

Viswa continued, "To me, the most powerful saying is by Leonardo da Vinci, 'Simplicity is the ultimate sophistication'."

"Sir, how far do you think leaders should exhibit simplicity?" Shyam asked.

"Good question," Viswa replied. "Leadership doesn't necessarily have to be about brand building. While there's a growing trend today to brand individuals and processes to project leadership, that doesn't define true leadership. In fact, large investments in personal branding can sometimes create myths around the individual, rather than building an authentic aura around the process or product—and without any real guarantee of impact. Leadership isn't about following trends or wearing the most expensive clothes. Many of our political leaders like Kamaraj, Lal Bahadur Shastri and others were role models of simplicity, living with very few possessions yet commanding deep respect."

"Are you saying corporate leadership requires some level of exhibitionism?" I asked.

Viswa laughed and said, "Well, exhibitionism may be a harsh word. They tend to acquire some unique branding. Again, there're good and bad examples everywhere. A good leader chooses what he wants, rather than what he is impressed with."

"Great. One of the key traits of a good leader is the ability to make the right choices; often the choices they make shape their own destiny and impact their followership," I said. "Do you think leaders always have to walk an extra mile?" asked Amar.

I responded, "Most successful people are those who are willing to walk that extra mile. It's not about competition, but about the drive to be better than the best."

"Napoleon Hill, a famous motivator and author, once said, 'You can start right where you stand and apply the habit of going the extra mile by rendering more service and better service than you are now being paid for'," I continued.

"Walking an extra mile is simply a conceptual advancement in any action. It adds value to one's leadership. It projects the leader in a better light. People tend to look at him curiously to draw motivation for their own selves. They tend to see him as an impressive role model," I added.

"When I see the farmers getting up early in the morning, almost at dawn, and walking to their fields with their tools, I'm inspired. Can I take that as barefoot leadership?" asked Amar.

Ankita smiled and said, "Why not? Don't we have a barefoot leader in Muni? Hasn't he inspired and motivated us, along with so many others in this village?"

"He has a limited following," said Shyam.

Viswa chimed in, "Let's take it to a more conceptual level: walking for a cause, walking with a purpose and walking for some human concern."

Amar noted, "There are many people with a missionary spirit who go to serve in the villages and remote areas of the country, just to serve a purpose and see their work as service to God".

"Do we need a huge following to be a leader? Is leadership determined by the number of likes or TRPs, like in the modern social media setup?" I asked.

"Social media has, in recent times, redefined the idea and meaning of followership," Viswa said. "There are many people with thousands, even lakhs, of followers, either because of their charisma or the momentary emotional impact they generate. But these followers aren't always genuinely aligned with what their so-called 'inspirators' stand for or actually do!"

"A good leader attracts people. They don't seek a following. A following is a result of their actions, performance and passion—the fragrance of which spreads throughout society," he added "They inspire, they motivate, and they lead their followers to a cause or a purpose," he said.

Just then, Muni diverted our attention.

"Sir, do you feel the soil? Do you feel the gentleness of Mother Earth?" asked Muni. I simply smiled.

I had never walked through such marshy land before—wet, raw and … I couldn't quite describe it. I just felt it.

"Not only should you feel it, but you should also smell the soil," Ankita remarked.

"Ankita, I agree. Smell is one of the earliest sensory inputs that foetuses experience in the mother's womb. In our lives, smell alerts us to an anticipated change. It tells us that something different is happening nearby and prepares us for what's to come," I said.

"True, Bala. I agree with you. A leader can only survive if they smell the change and are willing to respond to it. Not only that, but an effective leader is one who is willing to lead the change," Viswa responded.

"Agreed, Viswa. In a world constantly haunted by change, any organisation unwilling to adapt to emerging needs will face slow extinction. It will become outdated," I remarked.

"No wonder many top corporate companies, which once had roaring businesses, vanished because they didn't smell and respond to change," Viswa said.

"Sir, why do you say the world is constantly haunted by change?" Shyam asked me.

"Shyam, several decades ago, changes happened over a period. It took years for a change to manifest fully and complete its cycle. Communication across the world was slow, so even messages about change took time to spread. But now, with communication happening in nanoseconds, the flow of information is rapid and overwhelming. We can't afford to take our time to get ready and respond," I responded.

Muni was listening, possibly understanding the essence of the discussion.

"In agriculture too, there have been mind-boggling changes. Not only the tools and appliances have become modern, but the processes are so updated that actions that once took days are now done in hours," said Amar. "For example, we get real-time information about local weather, the growth and impact of insects, cattle diseases and immediate actions required to prevent plant diseases. It's truly awesome."

"But what about the cattle you still use?" I asked.

"They are our legacy. They represent our culture and reflect our cohabitation with the biosphere," said Viswa.

"Nevertheless, sir, our engagement with technology in agriculture is increasing every day. Many tasks that once required cattle are being done with technology. As Alvin Toffler writes in his book, *The Third Wave*, we moved from muscle power to machine power, from machine power to microchip power and now from microchip power to mind power. This shift is evident in the quality of our produce, our reduction in loss, and our waste management methods. We have no reason to regret our strategies in agriculture management," Shyam said.

"Of course. I'm reminded of the Green Revolution in the country—it was truly groundbreaking," I said. "It transformed our entire approach to agriculture and farm management, thanks to the visionary efforts of Bharat Ratna M.S. Swaminathan."

"And if we're talking about transformative movements," Shyam added, "we should also mention the White Revolution. Verghese Kurien's leadership in spearheading the cooperative movement in Anand, Gujarat, brought Amul into the spotlight and revolutionised the entire dairy industry."

"Both M.S. Swaminathan and Verghese Kurien were extraordinary role models in leadership," I said. "They touched the social consciousness not just at the individual level but across entire communities."

"In both cases, what stands out," Viswa added, "is their focus on generating leadership at the grassroots. If you look closely, these movements empowered people to access local markets and established a systematic, consistent and methodical supply of produce to maximise both its consumption and fiscal values."

Viswa looked at the young leaders and added, "You must read the biography of M.S. Swaminathan, popularly known as the 'Rice Man of India'. The challenges he faced as an agricultural scientist and his efforts to transform the country from a 'food-deficient nation' to a 'self-sufficient nation' were remarkable. It not only brought Indian agriculture into the limelight but also elevated the country's status in the global agricultural map."

"Interestingly, neither of these leaders craved authority, branding or recognition. Yet, name and fame followed them as a reflection of the work culture they demonstrated," he added.

"Did they lead from the back?" asked Amar.

Viswa responded, "They led from the front, from the back, from the side—they led by their presence and by their absence.

In short, in both cases, the leadership mattered more than the leaders themselves."

"I think farming is not just about land management," said Ankita. "It embraces a universe of related activities, each offering opportunities for different styles of growth, development and leadership."

Muni excused himself. "It's time to feed the cows and give them some water. They must be tired by now," he said. His compassion for the cattle reflected an inclusive approach to operations management.

"Managing cattle is critical in farming. Farmers need to take care of their health, as they are often infected by insect bites or diseases that require quick identification and immediate treatment," said Viswa.

"Unfortunately, they can't communicate," said Amar.

"No, they do communicate in their own way. We just don't understand their messages," replied Ankita.

"Ough, farming is not just about looking after the land. There are many related issues that are part of the process," I remarked.

"Farming also involves housing the cattle, giving them baths, providing feed and medicines, and of course, addressing issues in animal husbandry. And then there are the challenges in waste management," said Amar.

"Muni has extensive experience in this area. Sometimes, he can predict what disease the cattle have just by looking at them," observed Viswa.

"You learn by living with them," said Muni. "It's like a good leader understanding the pulse of the organisation. He understands the nuts and bolts of the organisation," he added.

"Understanding the pulse?" asked Shyam.

Muni smiled and, in a colloquial expression, said "Heartbeat". Shyam was doing some Google searching.

"Hey, what're you doing?" asked Ankita.

"Ankita, I was searching for a quote I read in a book called *The Unsettling of America: Culture and Agriculture*," Shyam replied. "I just found it. Wendell Berry, the author, says, 'The soil is the great connector of lives, the source and destination of all. It is the healer and restorer and resurrector, by which disease passes into health, age into youth, death into life. Without proper care for it, we can have no community, because without proper care for it, we can have no life'."

We all agreed that the words carried profound meaning.

> **REFLECT**
>
> 1. Trust is one of the most invaluable attributes of leadership.
> 2. Motherhood is an exemplary model of self-leadership dedicated to the well-being of others.
> 3. The simplicity of a leader is an asset that adds value to their profile and enhances their acceptability.
> 4. A true leader should be willing to go the extra mile to inspire their followers.
> 5. A competent leader understands the pulse and heartbeat of the organisation.

YOUR LEADERSHIP QUESTIONS

1. What are two significant initiatives you would take to enhance the level of trust within an organisation?

2. Mothers often exhibit exemplary leadership qualities, even in challenging circumstances. Identify and explain two major leadership traits you have observed in mothers.

3. Provide a few examples to explain what it means to "go the extra mile" and how doing so can be beneficial.

4. "I have a team of intelligent and talented individuals, but they lack the necessary competencies," said the head of an organisation. How would you help address this challenge?

5. Imagine you are the leader of an organisation and are under pressure to develop a personal brand. What would your focus areas be?

9
Leaders Walk the Talk

Become the kind of leader that people would follow voluntarily; even if you had no title or position.
—Brian Tracy
Canadian–American motivational speaker and author of self-development books

Walking alongside those farms was a refreshing experience. The day was getting warmer, but the feet were still cold on the wetlands. The three young people were more comfortable than the rest of us. Muni held his *dhoti* with one hand and had placed a towel over his head to shield his partly bald scalp from the sun. Viswa and I walked together.

"Muni, how many people showed up for work today?" Viswa asked.

"Only 12," Muni replied. He continued, "Three didn't turn up. Paru has a fever, Kala's son is unwell and Ramji sprained his leg when he fell off his bicycle."

I was surprised at how quickly Muni had all the details. Viswa had asked just one question, and Muni's response addressed all the potential queries.

"I think you've decentralised your leadership. Your team seems to have all the information you need, right at their fingertips," I said.

"Bala, a leader or an organisation may decentralise functions and delegate responsibilities, but unless things actually happen, there's no point in celebrating those functions on paper. It's important to delegate, authorise, assign responsibility, give ownership and then hold people accountable," Viswa explained.

"Yes, but how much should a leader intervene at each stage to ensure that things are being handled properly?" I asked.

"Sorry, a good leader doesn't do that," Viswa replied. "A good leader builds a nest of trust and creates a sense of pride in the middle-level leaders so that they believe they are movers of the organisation. This sense of pride becomes both the engine and the wheels of the organisation. In that sense, every employee, every team member is your PRO (Public Relations Officer) or an ambassador."

"That sounds good," I replied.

"In a true sense, the message of the organisation is delivered from the gatekeeper all the way up to the top man of the team," Viswa added.

"And does that mean even a servant can become a leader?" I asked.

"Ha, ha," Viswa laughed. "The idea of a servant is, in itself, an antithesis to the concept of a strong leadership. The hierarchical model oftentimes kills the spirit and joy of the employees, affecting their quality of work. However, there is a concept of servant leadership, which is different."

"Servant leadership?" Amar smiled.

"Sir, could you please repeat that?" Ankita asked.

"Servant leadership," Viswa repeated.

"What is it?" asked Shyam.

"It is a beautiful leadership style that seeks to move authority from management to the lower levels of the organisation, shifting the focus away from control over activities. It also differentiates power from authority," said Viswa.

"That's difficult to understand…." Amar responded.

"Okay, let me simplify it. It's a model that empowers employees with greater autonomy and fosters personal interaction among groups to build better synergy in relationships," Viswa explained.

"That sounds interesting," said Shyam. "But, sir, does the management or the leadership absolve itself from any direct engagement or responsibility?"

"No certainly not," Viswa continued. "This leadership style requires an individual to demonstrate qualities such as empathy, ownership, stewardship and commitment to the personal growth of others. According to Robert K. Greenleaf, the founder of the servant leadership concept and movement, 'The servant-leader is servant first … It begins with the natural feeling that one wants to serve, to serve first. Then conscious choice brings one to aspire to lead. That person is sharply different from one who is leader first'."

"Have we practised this earlier?" asked Amar.

Ankita said, "Plato, the ancient Greek philosopher, once stated that 'He who is not a good servant, will not be a good master'."

Viswa responded, "The concept of servant leadership is not alien to Bharat. We have several references to this in our epics."

"Can you give a few examples of servant leaders?" Amar asked.

"Of course, there are many—Swami Vivekananda, Mahatma Gandhi, Martin Luther King Jr., Nelson Mandela, Mother Teresa, Jack Welch, Cheryl Bachelder, Herb Kelleher and many others," Viswa said.

"Now I understand," Amar said.

Viswa continued, "Robert Greenleaf says, 'The first and most important choice a leader makes is the choice to serve, without which one's capacity to lead is severely limited'."

"Indeed, it's a great objective of excellent leadership. It's a model worth emulating," he added.

"Can this work in business and corporate world?" asked Shyam.

"Why not?" Viswa replied. "The concept is based on social consciousness and ethics. It can be, and should be, contextualised to all forms of business that profit from public services."

"I think it fits perfectly into the leadership that farmers provide," I said.

"True," they all acknowledged.

Muni was listening carefully. Viswa looked at him and said, "Muni, don't worry if you don't understand now. I'll explain all this to you when we sit together." Muni smiled. I saw in Viswa the kind of empathy required for someone working alongside him.

Ankita pulled a leaf from a nearby plant. It immediately curled up, withdrawing from the touch. "That's a touch-me-not plant," she said. "It is *Mimosa pudica*, the botanical name. In Hindi, call it *Lajvanti*."

"Just like a boss who withdraws from people the moment they want to reach out to him," said Amar. Everyone laughed.

"The boss syndrome is different from the leader syndrome," said Viswa.

"In which way, sir?" Ankita was curious to know.

"Well, a boss is overly conscious of his position and power. He enjoys a dictatorial approach and often reminds everyone that he can exercise his power over them. A leader doesn't do that," Viswa explained.

I intervened, "Let me share my experience with two real bosses I had.

"In the first case, I had just returned to the office after a fortnight's leave. I felt it was only courteous to meet my boss and inform him before resuming work. As I entered his room, I said, 'Sir, I'm back from my holiday.' He raised his eyebrows but didn't respond. I added, 'I just came to see you.' He looked at me and said abruptly, 'Yes, you've seen me—you can go.' I taken aback by his response; it completely froze me.

"In the second case, the boss was less qualified than I was and seemed quite conscious of it. Whenever I entered his room, his expression would tighten and I often felt like I was stepping into a lion's den. He had a habit of shouting at everyone, likely due to a sense of inferiority. This defensive attitude seemed to put him under constant, almost unbearable stress."

Everyone laughed. "The profile of the boss is getting outdated," said Shyam.

"By the way, did Ramji go to the hospital to see the doctor? If he needs medicines, I can order them for him," Viswa shifted the course of the discussion.

Muni responded, "He should have. Even if he's lazy, his wife would insist on his visit to the doctor." He smiled, and all of us did too.

We could see several women and men working in the field. Suddenly, Viswa stopped and asked Muni, "Isn't Latha on the family way?"

"Yes, sir," replied Muni.

"When is she expecting the baby? Is it safe for her to come to work during these days? Has she consulted a doctor?" One could read Viswa's concern on his face.

Muni went to call Latha. As Latha approached, Viswa asked, "Latha, how's your health? Has the doctor told you when you may go into labour?"

Latha, feeling shy, partly covered her face with her sari and said, "Sir, I'm in the last week of my seventh month. According to the doctors, it'll be another five or six weeks."

"Listen, Latha, you shouldn't be doing hard work now. This's your first pregnancy; you need to rest, do light work at home and take care of your health," Viswa advised.

"But, sir, I need to work. My husband doesn't have a job," she said, anxiety in her voice.

"Don't worry. You'll be paid. I'll give you your full salary for the next six months. Also, let Muni know about any medical expenses for the delivery, and I'll take care of it. But I want you to take care of your health," Viswa reassured her.

Latha wiped her tears. "Okay, sir, Thank you, sir."

Muni was also wiping his tearful eyes, moved by Viswa's compassion. He whispered into me, "No one does such things in this village. God bless him!"

Viswa looked at me and said, "It's no big deal. The government provides these benefits to all women employees, and most private sectors do too. However, these benefits are rarely extended to women in agriculture. Most of them are sent home empty-handed."

"Sir, is it because this is an unorganised sector?" Ankita asked.

"Yes. That raises the issue of the lack of leadership training in the unorganised sector. People aren't aware of the benefits they're entitled to through social security. Instead, someone with power and arrogance emerges as a leader, trying to control others. In many cases, it's a dynastic leadership," Viswa explained.

"Control, and not leadership, isn't it?" I asked.

"Yes, you're right," said Viswa. "There's also an element of exploitation in the unorganised sector. Leaders often keep their followers in the dark, preventing them from becoming educated about their rights and responsibilities."

"Though bonded labour has been legally eradicated, its mental attitudes still persist among some people," I remarked.

"Now, there are a number of government initiatives to help women," Amar added.

"But they often lack adequate information about these initiatives. Even when they do get benefits, they are treated like second-class citizens. What I'm trying to say is that concepts like gender equity, equality and neutrality have not adequately permeated the farming industry. For example, there's a gender-based payment disparity, which is illegal," Viswa explained.

Ankita smiled, clearly pleased with Viswa's words. "Sir, you are truly remarkable. The other day, I was looking into some statistics about women in agriculture. It's indeed interesting. The agricultural industry employs about 85 per cent of its workforce as women. Around 15 per cent of these women are self-employed."

I was impressed by Ankita's complete engagement with her subject.

She continued, "Unfortunately, out of the 85 per cent of women labourers working in agriculture, only 13 per cent own lands. The rest are just workers."

"Why is there such a high population of women workers in agriculture, from sowing to harvesting?" asked Amar.

"It's simple. There's a massive migration of male workers from rural areas to metro cities for other jobs," replied Ankita.

"That explains why we should be here!" Amar laughed.

"What about the scope for women in leadership positions in agriculture-related professions?" Shyam asked.

"I think it's increasing, and it's crucial. With more 'self-empowered' groups functioning in villages, there's an emerging opportunity for women to take on leadership roles. We need special professional training centres for women in leadership and management," Viswa said.

"Is that possible?" asked Amar.

"Nothing is impossible," Viswa said confidently. "You don't really need a Harvard degree to execute a local leadership. All you need is the power of your native language, common sense and the desire to lead."

"Sir, the other day I saw a wallpaper at one of my friend's houses. It showed two eagles flying, one over the other, with the statement, 'They do so because they think they can'. What you're saying fits perfectly with that," Ankita observed.

"But there's still no equity in payments for women, it is there?" asked Amar.

Viswa nodded, "That's true, and it's unethical. There should be no gender-based payment discrimination."

"What kind of leadership positions do you think women can occupy in the agricultural industry?" I asked Viswa.

"Sorry, your question itself is wrong. Tell me, which position can they not occupy?" Viswa responded. "We've had women as presidents, prime ministers and ministers in defence, finance and other ministries, as well as CEOs like Indira Nooyi. So, why not in all fields of agriculture? We shouldn't even consider the possibility that women can't lead in any field."

Ankita said excitedly, "You've hit the nail on the head, sir," and looked at me with a smile.

Viswa articulated the point more eloquently, "From space travel to space sciences, from film acting to film production, from advocacy to justice, from education to empowerment, from pilots to priests, women are everywhere. They should be and we should be proud of that. I'm confident, for example, that Ankita will make a big name as an agricultural leader, entrepreneur and innovator."

Ankita smiled and responded, "With your motivation and blessings, Sir. I believe the time is ripe to provide women with proper information regarding their constitutional rights, their privileges and the importance of their participation."

We walked past several men working in the fields. "Raju, how are you now? Are you taking regular injections?" Viswa asked.

"Yes, sir. I've had three injections so far, and I've been told there are two more to go," Raju replied.

Viswa turned to me and explained, "He was bitten by a dog, which is why he had to go through a series of injections."

Viswa then led us to a large well and showed us the water level. "Is the pump set working properly?" he asked Muni.

"Yes, sir, it's fine," Muni replied.

Turning to me, Muni said, "We had good rain this year, so there's no water shortage." He then brought a jug of water from the well and offered it to me. "Sir, have a drink and see how pure and tasty the groundwater is."

Viswa nodded thoughtfully, "Water management is a crucial issue in farming. We either have too much water or not enough. Both situations present challenges for farmers."

"Are there any innovations in water management?" I asked.

Shyam replied, "Yes, sir. There's increasing awareness around water conservation, storage, purification and its optimal use in agriculture to avoid wastage."

"We hold regular advocacy sessions and group meetings on these issues," Viswa added. "Experts from different organisations are invited, and all our farmers attend."

In a lighter vein, Ankita teased, "So, are the next few sessions going to be on 'Artificial Intelligence in Agriculture'?"

"Why not?" Viswa responded with a smile. "In some places, drones are already being used to inspect, oversee and monitor farms. Even fertiliser and pesticide spraying is being done with drones. Area-specific weather forecasting has also helped us make more targeted and successful efforts."

"I've heard that in some Western countries, robots are being used for harvesting," said Ankita.

"In spite of all these innovations, farming still has its own challenges," Amar remarked.

"Are we insured, sir?" Ankita asked.

"Yes, we are," Viswa replied. "But many farmers in this village don't fully understand how the insurance system works. I've asked the village *panchayat* president to organise a meeting where I can explain the nuances of farm insurance to them."

Muni turned to Viswa and asked, "Next week, sir?"

Viswa nodded.

We saw a middle-aged woman walking nearby with a load of crops on her head.

"Where is she going, Muni?" Viswa asked.

"Sir, to the eastern field," Muni replied. "It's time for crop rotation, you know."

"What are we planting now?" Viswa asked, curious.

"Chillies, sir. It's the best crop for rotation right now because it helps refresh the soil," Muni explained.

"Crop rotation?" I asked, intrigued.

"Yes, it helps in soil renewal. Further, farmers don't have to wait too long to plant again," Viswa said.

"It sounds like what happens in organisations when people are moved from one unit to another," I remarked.

"In organisations, we move people across domains for several reasons. First, they gain exposure to what happens in other units, helping them understand their challenges and develop an interdisciplinary way of thinking. Second, remaining in the same unit too long can lead to stagnation or overconfidence, causing their core skills to plateau. These rotations keep them engaged and adaptable. Lastly, leadership is a continuous learning process. To develop people as leaders, they must constantly learn from diverse sources," Viswa explained.

"By the way, how do you pay the farmers?" I asked.

"Through the bank," Viswa replied. "Their earnings are transferred to their accounts twice a month."

"Twice?" I asked, surprised.

"Yes, at the end of every fortnight," Viswa said.

"Don't you give cash to them?" I asked.

"No, not because I can't," Viswa said thoughtfully. "But I wanted to move away from the culture of someone holding power over their rights. When they draw their money from the bank, it becomes a privilege, not something given as a result of servitude."

"Amazing, you have crystal-clear thoughts on what you should do. Your unique approach is truly a different leadership style, I remarked"

"There's no particular style, Bala. Just honesty of purpose and truth. That's all. I believe this is what pays off in the long run," Viswa replied.

I turned to Ankita and said, "What's the difference between a manager and a leader? Every manager isn't a leader. A leader is a visionary. He doesn't just manage people; he leads them. And wherever necessary, he transforms systems and people."

"You mean a transformational leader?" Ankita asked.

"Every leader tries to bring about some transformation. But not all of them are transformational leaders," I explained.

"So, how do you identify a transformational leader?" Ankita asked.

"A transformational leader drives innovation, inspires new ways of thinking and harnesses a team's creativity to respond to change," I replied.

"That's a lot!" Ankita wondered.

"A transformational leader seems to be doing what a GM (Genetically Modified) seed does!" Shyam joked, and everyone laughed.

"What better example could we have than Viswa sir?" Ankita argued.

She might have been right. Viswa was indeed providing an exemplary role model to these young individuals. The youth of the country are undoubtedly looking for such mentors.

> **REFLECT**
>
> 1. Leadership hat focuses on serving people is always appreciated.
> 2. Servant leadership fosters participatory engagement aligned with a shared goal or purpose.
> 3. Compassion towards team members enhances a leader's respectability.
> 4. A true leader ensures gender equity, equality and neutrality among team members.
> 5. A transformational leader drives innovation and change.

YOUR LEADERSHIP QUESTIONS

1. How do you, as a leader, plan to connect with the bottom of your organisational pyramid?

2. Why is it important for a leader to "walk the talk"?

3. What are some visible roadblocks to achieving gender equity and equality in an organisation? As a transformational leader, how would you work to remove these obstacles?

4. What leadership attributes set a leader apart from a manager?

5. What are the limitations associated with the power of choice? How would you ensure its responsible use while being mindful of those limitations?

10
Farming and Leading in Crisis and Chaos

Optimism doesn't mean that you are blind to the reality of the situation. It means that you remain motivated to seek a solution to whatever problems arise.
—The Dalai Lama
Spiritual leader of Tibetan Buddhism

We all returned to the room for lunch. Muni had laid out the food on the table in a buffet style.

"Where do you get the food from?" I asked Viswa.

"Normally, I make my own food when I'm alone. But when I have guests, I have it organised by one of the families in the village. They cook and send the food to me for a very nominal cost. It's homely food, so I don't have concerns about the quality or safety," Viswa responded.

I replied, "Viswa, nowadays, food services have become a huge multi-crore industry in cities. With so many new delivery models emerging for domestic services, consumers can get all their needs delivered right to their doorsteps from their chosen vendors. When it comes to food, this has culturally impacted lifestyles and brought significant changes in family organisation and management."

"Similarly, leadership of new types have also emerged, with technology at the forefront, where you start leading people remotely, sometimes without even knowing their personal profiles or seeing their faces. I think the future holds promise for this kind of leaderships," I added.

"Yes. Some corporate companies sell things they don't own, others run transportation with vehicles they don't own, while some agencies sell food without actually cooking it, just acting as middlemen between consumers and producers. It's a new dimension of market engagement," Viswa responded.

"Sir, I've heard of the term 'Leadership in Absence'. Do these models relate to that?" Ankita asked Viswa.

"No, they don't," Viswa replied. "In that model, the views, actions, values and thoughts of leaders influence people by creating a sense of respect and value for what they've done. The business models we're talking about are remote distribution models that primarily focus on ensuring the supply chain meets the consumer's essential needs. Catering to their needs at their own doors creates a gravitation to those services because of enhanced comfort levels and saving time. Furthermore, with increasing challenges in travel, transport and the resulting stress, these models relieve the pressure at an acceptable cost. It's one way of exploiting opportunities in the service sector. There's a conscious effort to create a need for a product or service, and then drive people to see it as an unavoidable, essential need. Sometimes the focus is on creating an existential need, and sometimes, it's about creating a psychological need."

"Do you think this emerging model will expand to other domains of human activity, as long as it meets necessary gratification sooner rather than later?" I questioned.

"Well, there's is a new approach to marketing strategies these days. People talk about 'emotion marketing', which influences leadership across various fields, including marketing. This approach focuses on identifying and capitalising on opportunities to connect products and services with people's emotional needs and desires. It aims to create loyalty and engagement by continuously interacting with consumers. Patrick Dixon, author of *The Future of Almost Everything*, says, 'The future is about emotion: reactions to events are usually far more important than the events themselves'," Viswa replied.

He then looked at Ankita. "Sometimes, when these people come over, they prepare a meal. Ankita is an excellent cook. She has a keen eye for nutrition, and therefore, carefully selects balanced, healthy food."

"Eating nutritious food at the right time is crucial. Don't you think leaders need to look healthy?" Ankita quipped.

Everyone laughed. "It's not just about looking healthy but also having a healthy mind," Shyam said.

"You'll notice in many villages and on several farms, farmers will pause their work a few minutes before lunch, sit under a tree or whatever provides shade and eat with their family or friends. Then they take a brief rest before resuming work. Unfortunately, in many corporate houses, lunchtime has little to do with actually having lunch. One of the challenges in the corporate world is untimely and unhealthy eating habits," Viswa commented.

"Is that again a sweeping generalisation?" I asked.

"Perhaps. Let me rephrase it to 'several organisations'," Viswa responded. "In many businesses, employees deal with stress, anxiety, fear and conflicts the moment they walk into the office. And even if they have a lunch break, they simply grab something to eat in a rush, just to get it over with. They usually opt for fast food, and they don't even enjoy what they eat. Sometimes, they're not even aware of what they're eating. 'Over a coffee' has become an expression for sitting together to discuss matters, with a few cups of coffee or tea in hand, and it just continues from there."

"No wonder there's increasing evidence of health issues among people in their middle age, including heart attacks," I remarked.

Amar intervened, "Some companies, sir, organise annual health check-ups for their staff."

"Honestly, that doesn't help. You mismanage your health the entire year and then go for a check-up once during that period. How does that sound?" Ankita asked.

Viswa chuckled. "A good leader should be health-conscious, not only for themselves but for their employees too. No organisation can claim that health isn't their responsibility or priority. When you hire a person, you're also taking on their health concerns."

Ankita interjected, saying, "When asked what surprised him most about humanity, the Dalai Lama replied, 'Man! Because he sacrifices his health to make money, and then sacrifices money to recuperate his health'."

"True," said Viswa. "A good leader is a conscious health manager. He engages in preventive healthcare."

I added, "I've read the last words of Steve Jobs, the American investor who co-founded Apple Inc. and at the age of 56, died a billionaire. His words are worth reflecting upon. He said: 'I reached the pinnacle of success in the business world. In others' eyes, my life is the epitome of success. However, aside from work, I have little joy. In the end, my wealth is only a fact of life that I'm accustomed to. At this moment, lying on my bed and recalling my life, I realise that all the recognition and wealth that I took so much pride in have paled and become meaningless in the face of impending death. You can employ someone to drive the car for you, make money for you, but you cannot have someone bear your sickness for you. Material things lost can be found. But there is one thing that can never be found when it's lost – Life'."

Amar observed, "Sir, I was looking at the websites of several organisations and found that many of them have fitness centres, yoga classes, and healthcare clinics to address the health concerns of their staff. How do you respond to that?"

"You're right, Amar. I'm aware of that. But the real issue is how much those facilities are actually utilised. While they do showcase the organisation's intent, in many cases, the utilisation levels are low. Even when employees engage with these services, it's often superficial—mostly to give the impression that they're participating. Genuine, wholehearted engagement with health remains quite low," I said.

Viswa smiled. "I tend to agree. There's little real advocacy on health issues."

"Occupational hazards also pose a problem," Amar added.

The lunch on the table consisted of salad, cooked vegetables, wheat *chappati*s, *dal*, greens, curd and pickle.

"Light and good enough for a working day," said Viswa.

"Eating is not only a culture but there's also a science to how we eat," Viswa remarked. "In city life, I've seen many of my friends having a quick bite—sometimes standing, walking, talking or even while dressing. For many, eating has become just another task, oftentimes a burdensome one."

All of us laughed.

Viswa continued, "There's an interesting quote attributed to Peter the Great, former emperor of Russia, and I fully endorse it. He said, 'Destiny may ride with us today, but there's no reason for it to interfere with lunch'."

Everyone nodded in agreement, acknowledging the power of this statement.

Viswa further said, "The concept of large families was indeed a boon for nurturing the health of everyone in a collective and collaborative manner. The grandmother or an aunt at home provided the much-needed leadership and guided everyone on their health issues. A variety of dishes were prepared to cater to personal needs and preferences. Leadership at home coexisted harmoniously with leadership on the farms in our villages. These two forms of leadership complemented each other. People lived long lives. Healthy food, combined with hard physical work, helped them manage their lives peacefully and successfully."

"We've now transitioned to models of packaged, processed and preserved food consumption. In many cases, food is prepared in copious quantities, stored for extended periods using preservatives and then consumed not as fresh food but as 'refreshed' food. The concept of fresh food is slowly disappearing. Parents of yore played a significant role in ensuring their family members had timely meals. But now, with most senior members of the family seeking refuge in old age homes, these traditions are changing," I reflected.

Ankita observed, "I saw an advertisement from a real estate agent promoting apartments without a kitchen, suggesting meals could be easily obtained from a central kitchen. They even offered a monthly menu order serviced by a nearby food company."

"As we discussed earlier, consumerist tendencies are reshaping our lives and influencing the leadership models that navigate our thought processes," Viswa said.

"But, sir, the Covid-19 scenario changed our working environment a lot," Amar pointed out.

"Yes, both in good and bad ways," Shyam agreed.

"Right, it's really about how we view and perceive situations and how we choose to adapt them," I added.

"Sir, was Covid-19 a blessing or a curse?" Shyam quipped.

Ankita grew agitated. "Blessing? Don't even say that! It was a global calamity!"

"True. Besides the millions of lives it claimed, the chaos it brought to social harmony and the human suffering it caused can never be fully understood or expressed in words," I tried to intervene.

"That brings us to another interesting discussion: leadership in chaos," Viswa said.

"Sir, how do we differentiate a crisis from a chaotic occurrence?' Ankita asked.

Viswa replied, "Chaos is a state of utter confusion, disorder or the absence of a system or process in an organisation. A crisis, on the other hand, is a situation of extreme instability or danger that requires immediate attention."

"Can I say, just to make it light-hearted, that the difference is like comparing a cyclone to a tsunami?" I spoke.

Everyone looked at me. I added, "In a cyclone, there is an eye at the centre, so we know where it's going to land. By assessing its direction, speed and mobility, we can take preparatory steps to minimise the impact. On the other hand, a tsunami, even if we know it's coming, brings a level of confusion and calamity that can't be predicted. Its magnitude and intensity are exceedingly hard to gauge."

"Leading in a crisis calls for different skills than leading in chaos," Ankita smiled and asked, "Should a leader consciously walk through chaos, fully aware of the challenges ahead?"

"Certainly," Viswa replied, "If one doesn't, one lacks courage. Tom Peters, who authored the book *Thriving Through Chaos: Handbook for a Management Revolution*, said, 'Unless you walk out into the unknown, the odds of making a profound difference in your life are pretty low'."

"Isn't it difficult to walk through as a leader during crisis or chaos?" Amar questioned.

"Yes, it is, but that's where you reveal your true mettle. During the Second World War, Winston Churchill, the then prime minister of the United Kingdom, made a powerful statement: 'If you're going through hell, keep going.' That's what makes a leader tough," Viswa replied.

"The Covid-19 pandemic taught us several lessons and gave me two important messages," I said. "One, nothing matters in life; and two, everything matters in life."

"Wow! You've captured it very well," said Shyam.

"During Covid-19, we saw several models of leadership emerge," said Viswa. "Essentially most of them took us beyond materialistic definitions of leadership to more broad-based, socially conscious, humanistic and spiritual parameters."

"Sir, does leadership during a crisis or chaos create a brand for the leader?" Ankita asked.

"That's a sensible question," replied Viswa. "A leader may not seek to create a brand through their actions, nor should they exploit the situation to do so. Such attempts will backfire and reflect poorly on them sooner or later. But a leader's actions will naturally determine the kind of branding they will get."

"Maybe it adds to their profile," said Amar.

"Ha, ha! A true leader never drafts their reputation – others write their profile," Viswa laughed. "There is a saying, 'It takes a whole village to make a leader'."

At that moment, Muni walked towards Viswa and spoke to him. Viswa replied in Tamil, "Please ask him to come here. Get two chairs."

A gentleman, dressed decently in a white shirt and *dhoti*, sporting an *angavastram* (a traditional handspan cotton cloth worn on the shoulders) walked in with his wife, who was wearing a silk sari.

"*Vanakkam*," they greeted us in Tamil. . "Happy to see you, sir. We've heard a lot about you." Viswa stood up and welcomed them with folded hands.

"Thank you. But I'm too small in stature compared to legends like you who have provided leadership for several villages in this area," Viswa said with humility.

The gentleman smiled and replied, "That's all history now. Today, I'm like an extinct species because people respect you only when you have money or power, when you're useful to them and until they can gain some advantage from you. I did a lot, and I also spent a lot. I helped many people and many organisations. But as you know, I was cheated by some. They took advantage of my goodness, and at some point, I started questioning myself—whether being too good is too bad." He laughed and said, "I've lost everything."

"Sir, we often fall and are injured, but that doesn't mean we have failed. There's a powerful quote by Anglo-Irish author Oliver Goldsmith: 'The greatest glory in living lies not in never falling, but rising every time we fall'. Even Edmund Hillary did not conquer Mount Everest on his first attempt," Viswa said.

"I agree. But you need age, energy and the moral strength to strive. I have lost that," the gentleman replied.

"Please don't say that. I see you as a treasure trove of energy and wisdom. You are one of those rare leaders who remain leaders throughout their lives, whether or not others acknowledge it," Viswa said.

Ramaswamy Mudaliar, as he is popularly known, smiled, and said, "Let me tell you the purpose of my visit. I still have about 20 acres of fertile land left, down from the 400 acres I once owned. I thought I must hand over this land to someone who can cultivate it as a farm, and I'm willing to accept any reasonable market price."

"Selling? Oh, no," Viswa argued. "You must use and cultivate it yourself. Return to farming, sir. Perhaps the golden days will return. Mother Earth is gracious and blissful. And your commitment and honesty are well known."

"I have discussed this with my family and made this decision," Mudaliar emphasised.

"But sir, you have a larger family—the family of farmers who still respect you and see you as their god. You shouldn't disappoint them," Viswa said.

"To be honest," Mudaliar said, his voice tinged with emotion, "I'm surprised by your advice, and I'm thankful. You're the only one who has asked me to return to farming, and I appreciate it. Others have declared me a massive failure, including members of my own family." His eyes were wet. "My sons have migrated to cities and are doing business there, claiming they earn more profits. Perhaps they do, earning more money with less labour."

"Sir, don't worry. I will stand behind you and give you all the help you need. I can buy that land, but for what? If I have a friend who leads the community with grace and grandeur, I will consider myself truly blessed," Viswa said.

Mudaliar looked at his wife. "Even she thinks that I must call it a day and settle down."

"Sir, you're only in your mid-50s. This is no age to retire. This is the time when your wisdom and experience should come into play. No doubt, you may have become a bit too ambitious or went adrift for a while. As a result, you lost your land and money to a competitor. But there's a saying by Hellen Keller: 'A bend in the road is not the end of the road, unless you fail to make the turn'. Human life is much like a journey on a winding road. No one ever experiences a smooth, 'problem-free' journey. The key is how we adapt to change, adjust our course, and have faith that better things are yet to come. You can restart your life now. There is boundless joy in rediscovering yourself and showing the world that we are always winners," Viswa responded.

"Your farming techniques and approaches are new. How can I respond to these changes?" Mudaliar asked.

"Sir, I entered farming as a novice just six years ago. Your right-hand man, Muni, became my *guru*," Viswa said.

Mudaliar looked at Muni, who nearly began sobbing.

"We all learn throughout life. Often, we learn through our mistakes. So, don't worry. Be strong. Visit me when you're free, and I'll help you draft a plan of action," Viswa said.

"But I don't have much money for investments," Mudaliar replied.

"Don't worry, sir. I can help you get an agricultural loan, and I'll stand as the guarantor. Where there's a will, there's a way," Viswa said.

Mudaliar stood up with renewed confidence. "Brother, you're younger than me, or else I'd touch your feet. I rarely find any leader in the community who genuinely feels happy for another person's growth and well-being. God bless you. I'll meet you in about a week."

As he walked away, his wife wiped her eyes and followed him. Muni followed them to the gate.

I sat in silence for a while, and then said, "Viswa, you're great".

"Not at all," Viswa said. "I've just given him some motivation, some encouragement and helped him realise his own strength, because I've heard a lot about him."

"What happened to him, sir?" asked Ankita.

"Well, he was once the undisputed leader of the farmlands here. His leadership was driven by money and authority, not by intellect and pragmatism. But people took advantage of him–he was exploited and lost a major part of his wealth," Viswa replied.

"You place a lot of trust in people," I remarked.

Viswa replied, "Bala, I recently watched an interview of Sundar Pichai, the CEO of Alphabet and Google. When asked about his role as a leader, his response was quite impressive. He said, 'As a leader, my job is to make other leaders in the organisation successful. My job is to remove any roadblocks that might stand in their way'. I think that's a valuable lesson for us all. By helping those around us succeed, we, too, achieve remarkable success as leaders."

"Bala, you might also have heard the famous saying by Oscar Wilde, the Irish author: 'Every saint has a past, and every sinner has a future'. I see a bright future in Mudaliar's eyes. Remember, a good leader is one who grows other leaders. In his case, all we need to do is help him rediscover his own potential," he added.

Muni returned and asked, "Shall I make some coffee?"

"No, I'll make the coffee," Ankita said as she moved towards the kitchen.

> **REFLECT**
>
> 1. The physical, emotional and psychological health of a leader is a strong testament to their leadership for their followers.
> 2. Crisis management skills enhance the quality of leadership.
> 3. A leader's ability to handle chaos and restore order reflects a pragmatism grounded in wisdom.
> 4. A leader should not only be resilient but also empower followers to become equally irrepressible.
> 5. A true leader neither exploits the weaknesses of competitors nor capitalises on their failures.

YOUR LEADERSHIP QUESTIONS

1. Reflect on how you handled a crisis on two different occasions. What skills did you use at the time, and how would you approach those situations if they occurred now?

2. Describe a time when you lost hope in your leadership, then became resilient and fought back to succeed. What changed for you?

3. Identify one significant failure in your professional journey that you turned into a stepping stone to success.

4. If a partner in your organisation let you down, what two different approaches would you take to handle the situation?

5. If you made a wrong decision despite others' advice and later realised it, what steps would you take as a leader to restore your credibility in the organisation?

11
Braving the Storms: The Farmer and the Leaders

Life isn't about waiting for the storm to pass ...
It's about learning to dance in the rain.
—**Vivien Greene**
British writer

Muni had left for his home in the evening. Ankita cooked dinner with the assistance of Amar and Shyam. As we sat for dinner around 8 p.m., we noticed that dark clouds had begun to gather.

"It could be a thunderstorm," said Viswa.

"There's thunder and intense lightning," Amar pointed out.

"Does it rain this time of year?" I asked.

"Well, these are summer thunderstorms. They occur due to temperature conditions that lead to local disturbances," Viswa explained.

"We have an agricultural information centre as well as a weather forecast centre in the city. It's about a two-hour drive from here," Ankita informed me.

"Well, technology has come a long way, providing information and even offering advice on precautionary and operational strategies to deal with such eventualities," Viswa said.

Rain began pouring heavily, accompanied by strong winds. The howling of the wind and the rattling of windows and doors could be heard. A few dogs were barking in the night.

"Is this rain good for the crops?" I asked.

"If it's moderate, yes. If it's in excessive, no," Viswa replied.

"Sir, should I take the umbrella and go to the cattle house to check if they are all right or if they need any support?" asked Shyam.

"Don't worry, Shyam. Sethu is there; he'll take care of them. He told me he's going to stay there overnight," said Viswa.

"Does waterlogging in the fields cause some anxiety?" I asked, curious.

"For the current cycle of crops, it's not a problem, but you can't say the same when crops get rotated," Viswa responded.

I commented, "It's like saying that one set of leadership concepts may be relevant to a particular situation, but not necessarily to all. Even if there's a chance to apply those concepts contextually, one must be prudent and use a lot of common sense."

"Back to leadership, is it?" Viswa said sarcastically.

"I must give you a piece of advice, Bala," Viswa said, looking at me. "When you talked to me before coming to the village, you had a biased view. You probably felt that farmers and villages largely depended on the wisdom of modern leaders. You couldn't have imagined that a lot of traits of excellent leadership could be found here, even though people didn't have formal education. I'm just trying to remove the myth that a degree or certificate is a gateway to good and effective leadership. A good leader is one who can manage his mind and thoughts as needed, not simply by always applying rules and policies. The leaders here rely a lot on gut feeling and intuition."

"Are you saying that qualifications don't add value to leadership?" I asked.

"I didn't say that. The spirit of leadership is a calling of one's soul. Other factors can add value, but being human, we need to be in harmony with the environment. One major cause of stress is 'living out of place'. Carrying professional challenges home and thinking about domestic issues at work creates a psychological imbalance. Your focus, attention and energy get dissipated," Viswa said.

"It's difficult," I replied.

"Difficult, but not impossible," Viswa responded.

"Sir, I have a question. Aren't you worried about the impact this rain will have on your crops? It could pose a threat or lead to a loss," Amar asked Viswa.

"You're right, Amar. The thunderstorm will have some impact, but I don't know to what extent. Will it be good or bad? More or less? Where and how much? Without adequate data and information, all we're doing is worrying and suffering from anxiety and anticipatory fear," Viswa replied.

"Isn't that an exercise in problem-solving?" Ankita asked.

"No," Viswa replied. "In any crisis, we shouldn't panic. We must avoid reacting out of fear and instead focus on responding thoughtfully. The true quality of leadership is revealed in how a leader responds."

He continued, "It's a futile exercise otherwise. Effective problem-solving must be grounded in data and reliable, credible information. Without that, it becomes mere speculation and that often leads to gossip. A leader not only needs accurate data but also a deep understanding of the parameters they operate within: the key influencers, the obstacles to resolution, the necessary resources, and a clear methodology for solving the problem. Sometimes, challenges shape us. As Charles Caleb Colton, the British writer and cleric, said, 'Times of great calamity and confusion have been productive for the greatest minds. The purest ore is produced from the hottest furnace, and the brightest thunderbolt is elicited from the darkest storms'."

"Farming provides the opportunity to gain experience in leadership at every stage," said Amar.

I raised my eyebrows.

"Yes sir, it teaches you to deal with the present rather than the past or the future," Amar added.

"Can you elaborate on that?" I asked.

"You see, one must be watchful and vigilant about the growth profile of the crops almost every day. Not only storms but also insects, rodents and other infections carried by the winds can have an impact on the crops," Amar said.

"The farmer has to play the role of a policeman, always vigilant, until everything settles down," he added.

"I think every village needs to have a vigilance department to counsel farmers on these issues," I remarked.

"Of course, sir. The agricultural department of the state, for example, has a monitoring cell on such issues," Amar said.

"The real problem now is that the food is getting cold," Ankita smiled.

The team brought the dishes to the table.

"Thank you, guys," said Viswa, then he looked at me.

I smiled and said, "Thank you, guys!"

Through the night, the rain became heavier, lashing the area. One could see, or rather feel, how a thunderstorm makes a strong impact. Along with the thunder, we could hear the barking and howling of the street dogs. I saw Viswa getting up and opening the door to let two of the dogs inside. I understood that compassion doesn't need any certification.

In the early hours, the rain stopped. Viswa went and sat on the granite bench outside the house. A few minutes later, I heard some conversation. A few villagers had gathered near Viswa.

"I think my crops have been badly affected," said one villager.

"A couple of trees have also fallen down," another villager added.

"I expect that many plantain trees must have fallen, because they can't withstand these winds," said yet another.

There were a lot of anxieties expressed, and Viswa sat silently. Then he said, "No wild guesses, please. Let's wait and see after the sunrise. Don't jump to conclusions."

Ankita made some filter coffee for all of us.

"Did you stay here overnight?" I asked her.

"Yes sir, all three of us did. You know we have two rooms on the first floor. Amar and Shyam took one, and I took the other. I had informed my parents that I would stay overnight. They are simply fascinated by Viswa sir and often tell me that I will never get a teacher like him in my life," Ankita replied.

"Sir, have you seen snails outside?" she asked.

I accompanied her to the backyard and found scores of snails, big and small, and some frogs.

"I was reading an article on the internet titled 'Leadership Lessons from Snails: Creating Safe Spaces' by Dr Robert Osinbajo in which he writes, 'A snail feels safest in its shell. The environment has to be right for it to come out. Similarly, leaders must create a safe environment that encourages employees to step out of their comfort zones'."

"Indeed, a wise leader facilitates followers to come out of their comfort zones but ensures that they don't end up in a mess because of an insecure environment," I replied.

"Is this approach correct? Is it not wise to prepare them to manage challenges by letting them face possible situations before they are confronted with the reality?" asked Ankita.

"Yes, they need to, but leaders should not let their followers step into risky and unsafe areas just to satisfy their own egos. A leader should not make their followers sacrificial goats," I said.

"Sir, you seem to be overreacting to that statement," Ankita said.

"Yes, you're right, Ankita. I was wondering how many followers are made sacrificial goats, even in politics," I replied.

"It happens even in business and corporate ventures when someone becomes a scapegoat for things they are not at all related to. A leader should not be an escapist but should be able and willing to accept responsibility for situations that occur under his stewardship," Viswa said.

"Normally, in any team, two types of games are played: the first is passing the buck, and the second is escapism," I observed.

"A leader who passes the buck to a team member without taking any moral responsibility will soon be exposed, putting the leader in a poor light," Viswa said.

"In general, when there is appreciation, we want to be at the forefront, but when it comes to criticism, we tend to distance ourselves from it. An effective leader does not behave that way," I commented.

"Some leaders say, 'They do their job, and I do mine,' while others say, 'They are employees; I am also an employee.' Then there are those who say, 'They do their job, and I intend to get it done,'" Amar replied.

"True. There's the 'I ,I' attitude, 'I, You' attitude and the 'I, we' attitude," Ankita commented.

"Among the farmers, there is no question of passing the buck. Everyone takes responsibility for what has happened to him or his farm," Shyam said.

"Possibly, there is no other option," he laughed.

"Here comes Muni. Shall we go together and find out the post-storm scenario?" he asked.

"Yes. Let's go," said Viswa.

On the way, Viswa stopped outside a small house with a thatched roof, where an old woman was sitting. He immediately turned to her and asked, "*Patti* (grandma), how was the rain last night?"

"Viswa, my grandson tells me that the crops are under water. The water must be drained out soon," she said.

"Okay, ask him to talk to me if he needs help to hire a pump. He must know what to do, but still, I'm here for any help," Viswa said.

"Thanks, *da*....," the old woman responded.

After a short distance, Amar said, "Sir, you know that her grandson is not very competent. You could have helped her directly."

"Sorry, Amar. Never underestimate the competency of others. A leader shouldn't. Even if they have difficulties in doing something, help them to learn how to do it. One should find out what they are capable of. A leader should not create a sense of dependency among their followers. Enabling the team for self-learning and self-directed learning is critical to good leadership," Viswa replied.

After a few steps, Viswa stopped again. "Hey, Pandi, what happened? Any damage last night?"

Pandi responded in local colloquial language, "*Thalaiva* (leader), about 100 plantain trees have fallen. I don't know what do."

"Listen, what is gone is gone, but I will take some photos and send them to you on your mobile. Claim compensation from the insurance company. By the way, harvest the vegetables and fruits that are unharmed so that they can be sent to the market.

Otherwise, at least send them to the cattle house so that they can be used as feed for our cows," Viswa said.

Ankita smiled and said, "Lateral thinking?"

"Viswa, as a leader, has great emotional intelligence. He reaches out to everyone. He empathises with all happenings and situations. He listens to the pain of everybody and feels it. He is willing to solve the problems of others with commitment and provides alternative solutions. He goes out to help people in a financial crisis but has no expectation from others," Muni spoke in Tamil in a low voice.

In a few minutes, at least eight people gathered around Viswa. He stood at the corner of the street, listening to their tales of sorrow. He proved that a good leader is a great listener. At one point, he stopped them and said, "Listen, we all have problems, one kind or another. If we put our heads, hearts and hands together, we can solve all of them. Collaboration and cooperation help us to see win-win situations."

Everyone nodded in agreement.

"There should be no priorities, no pushing or pulling, no gossip, no misgivings. We should trust that each of us is serving the other. All we need is 'many hands and one mind'," he added.

Amar questioned, "Sir, why one mind? Doesn't that curb diverse views and opinions?"

"Well, I should have been clearer in my communication. What I meant was a shared vision," Viswa said.

I stood in admiration. How had Viswa managed to develop the concept of a "shared vision" among these farmers? How had he won their hearts and brought them together as one big team?

As we continued walking together, in a relaxed manner, I shared a short anecdote I had read somewhere, "A young man once went to a concert by a famous orchestra. The programme was amazing. When he was returning, he met an older, senior musician on the way. To impress him, this young man said, 'Sir, I just returned from that amazing orchestra concert.' The old musician smiled and asked, 'By the way, did they play together, or did they play simultaneously?'"

Everyone looked at me.

I continued, "People may work simultaneously, but that doesn't mean they work together. In the first case, there's energy; in the latter, there's synergy."

"Just like the difference between noise and music," Viswa added.

Amar responded, "True. A great lesson in teamwork."

"A team is no team if they don't collaborate," I said.

"Sir, you are indeed a lion in leadership," Ankita remarked, pointing at Viswa.

"No, I'm just a sheep," Viswa smiled.

I intervened, "There is a saying attributed to Alexander the Great: 'An army of sheep led by a lion is better than an army of lions led by a sheep.'"

"What would you say if you had just a herd of lions?" Viswa's words reflected the trust and confidence he had in the people.

Ankita smiled and said, "Sir, in the book *Animal Farm*, George Orwell says 'All animals are equal, but some animals are more equal than others.'"

Viswa joked, "So, Ankita, you're determined to classify me in the category of animals, huh?"

We couldn't control our laughter.

REFLECT

1. In a crisis, leaders should respond thoughtfully, not react impulsively.

2. A leader who avoids responsibility and shifts blame is seen in a poor light.

3. One of the major responsibilities of a leader is to develop a shared vision.

4. A good leader listens patiently and actively to their team members' concerns.

5. A leader works *with* the team, not *on* the team.

YOUR LEADERSHIP QUESTIONS

1. What are the disadvantages of reacting to a situation instead of responding thoughtfully?

2. What causes team members to pass the buck? What should a leader do when things go wrong?

3. What steps would you take to develop a shared vision within your team? How does a shared vision benefit an organisation?

4. What is the difference between hearing and active listening? In what ways does active listening support effective leadership?

5. How does working with a team offer more support to a leader compared to working on a team?

12
Problem-Solving on the Riverside

There are not more than five musical notes, yet the combinations of these five give rise to more melodies than can ever be heard. There are not more than five primary colours, yet in combination they produce more hues than can ever been seen. There are not more than five cardinal tastes, yet combinations of them yield more flavours than can ever be tasted.

—Sun Tzu
Ancient Chinese military general, strategist and author of The Art of War

Around 4 p.m., Ankita suggested, "Let's take Bala sir for a walk to the riverside." I was happy to join.

"You might have to walk about 2 km to reach the river," Amar mentioned.

"That's fine. I missed my morning walk, so this will make up for it," I replied.

"Sorry, sir, but the objectives of a morning walk and this evening walk are different," Ankita said, as Viswa nodded in agreement.

"By the way, Ankita," I asked, "as someone interested in English literature, have you read the essay 'The Walking Tours' by R.L. Stevenson?"

"No sir," she said.

"It's an essay worth reading. I studied it in college, and since then, I've read it more than a hundred times. Stevenson explains how one should walk, especially in different places, at different times, on different occasions and with different people," I told Ankita.

"Oh my God, that sounds interesting! You're right. Every type of walk has a different purpose or aim. And the mood and energy levels change depending on the time, place and company," she responded.

Viswa said, "Walking is very important for leaders because it helps open up the mind to free thought. It's part of body language training. In many cases, leaders are trained on how they should approach the stage, how far they should stand from others and the right way to position them. This is critical in relationship management."

"Talking while walking can help share concerns and views. It benefits some people in this way," I added.

"The management jargon 'Walk the talk' is quite popular," Amar observed.

"Yes, a great leader does walk the talk. But the deeper meaning is about his personal value system, which is why people respect him," Viswa commented.

Just then, we heard a greeting: "*Vanakkam*, sir."

"Oh, Ezhil, how are you?" We saw Viswa speaking to a teenage girl with a backpack. Ankita greeted her warmly, holding her hand like a friend.

"How are your music and dance classes going?" Viswa asked.

"Sir, music is fine, but I'm planning to give up dance classes," Ezhil replied.

"Why? Have you lost interest?" Viswa asked.

"No sir, I'm still very interested. But my family can't afford the extra expense every month. As you know, my elder brother is in an engineering college, and Appa, being a farmer, doesn't have a regular income," Ezhil said.

"Absolutely not. You shouldn't stop for that reason. I'll talk to your father Siva. Ankita, please take the money from me and pay her one year's worth of dance fees on my behalf. I'll speak to her father about it," Viswa said.

"No, sir, I can't accept that," Ezhil told him.

"Listen, Ezhil. I know what's best. If you've an interest in learning something, you should follow it. This money is very little for me to give, and I'm happy to do so. But once you're ready to do a dance recital, you must invite me to one of your performances, okay? I'll see you later."

Ezhil smiled, holding Ankita's hands as they walked together with us.

Shyam whispered, "How many people will he help?"

Viswa overheard and replied, "You're right, Shyam. In leadership, you might solve a problem, but there's no guarantee that the same problem won't arise again. You need to address the root cause and find a lasting solution. You can't expect to solve a problem once and think a second problem won't come up, or that you'll automatically have the solution for it."

"Is there any other way to solve this problem?" asked Amar.

"Are you willing to wear your thinking hat?" Ankita joked.

"Yes, there is a solution. What if we ask the dance teacher to conduct classes in our community hall at a set time, and the *panchayat* pays her a salary as part of a welfare initiative? That way, it would help many others to join too," Amar responded.

"That's a good idea, but we'd need to speak with the *panchayat* and convince them first," Shyam said.

Ezhil thanked Viswa and left with a smile.

As we walked, I was once again touched by Viswa's gesture. Before I could speak, he volunteered, "One issue in Indian families is that when something has to be taken away, it's often the girls who suffer, not the boys. Our families need to learn to treat boys and girls equally."

I loved Viswa's statement. Ankita smiled and exchanged glances with her friends.

"Muni, can we stop for a few moments at our cattle house?" Viswa asked.

"Yes, sir," Muni replied, and led us to the cattle house.

Curious, I asked, "Should it be a 'cattle house' or a 'cattle shed'?"

Viswa smiled and explained, "I call it a 'cattle house' because the moment you say 'shed', it suggests a second-rate, inferior approach. The word 'house' gives them the respect they deserve. They do so much for us, and I think they deserve that respect."

"You're right. All our problems begin with our thoughts," said Amar.

"Yes, including wars," replied Viswa.

We entered the cattle house, which was neat and clean. The cows had their food placed in containers, and about four people were working there to care for them.

"Mohan," Viswa called out to one of workers, "How is Shoba doing now?" He was referring to a cow that had recently delivered a calf.

Mohan replied, "She is doing fine, sir, along with her little one. She's slowly gaining weight and she should be fit soon."

"Good. Don't put her under any stress. She needs to spend a lot of time with her calf to keep her emotions calm," Viswa advised.

I listened patiently to the conversation, surprised by how Viswa, as a leader, seemed to know every single detail in his operational universe. Moreover, his attention to animal welfare also impressed me.

Viswa spoke, "A good leader is like a skilled navigator. The navigator needs to know about every nut and bolt of his vehicle, ensuring that every screw in the engine is functional."

Ankita and the two young boys had gone inside the house and were almost conversing with each cow. When they called them by name, the cows responded positively.

"It seems to me that all of you are touching lives," I said.

"And animal's lives too, sir," Shyam added.

"You're right. Good leadership is about touching lives all around," I replied.

Amar turned to Viswa and said, "Sir, I think Booma is unwell. She's not responding to me like she usually does."

Muni confirmed, "Amar is right, sir. Booma seems to have a foot infection. She's unable to stand comfortably."

"Did the doctor visit today?" Viswa asked Muni.

"Sir, he came yesterday. He's gone home today because his sister is getting engaged," Muni replied.

I was surprised at how Muni could know every detail, even about the doctor's family.

Viswa seemed to anticipate the question in my mind and said, "Dr Hari treats Muni like his elder brother. He has a lot of respect for Muni. Even when he finds it hard to control some of these cows, he asks Muni for help. Muni shared a personal, family-like relationship with each of these cows."

Muni called out, "Shanti, come here," and in a few seconds, the cow named Shanti slowly walked towards him.

"That's the emotional bond anyone should have with the team they manage or lead, even when it's with non-human beings," Viswa remarked.

We continued walking toward the river. A little further along, we saw four to five people sitting under a Pepel tree. When they saw Viswa, all of them stood up.

"*Vanakkam*. Please sit down. You shouldn't stand for me— I'm much younger than all of you," Viswa said.

One of them replied, "*Thambi*, a person is respected not for his age, but for his heart and his service." Muni nodded his head like an elephant would do, and all three disciples laughed.

"Anything going on?" Viswa asked.

"Sir, in the next two months, we'll be planning the village temple festival. Once we've finished our discussions, we'll come to you for your final advice," one of the men said.

Viswa smiled and replied, "I'm too inexperienced for all of this. You have years of experience, so I'm not sure I can contribute much."

One of the men told me, "Last year, he gave us so much guidance on the safety aspects of the festival. Thanks to his planning and advice, we avoided a major fire."

We continued walking.

"Sir, what's the most critical aspect of any planning?" Ankita asked.

"Planning should be a collaborative activity, where both heads and hearts meet. It's where ideas and strategies align, where logic and objectives converge. Above all, there must be a convergence of perspectives," said Viswa

"How pragmatic should planning be?" Shyam asked.

"If it's not pragmatic, it's not really a plan—it's something else," Viswa laughed.

"Sir, how do strategies differ from plans?" Ankita sought to clear her doubt.

"Well, I'm not a professor of management or leadership, so I can't give a professional answer. But from my perspective, many people plan with only their desired outcomes in mind. Some focus solely on achievement goals. The problem is, they often overlook potential roadblocks, leaving them confused when challenges arise. Strategies, on the other hand, are grounded in logic. They involve operational steps to solve problems and offer alternative paths forward," Viswa said.

"As such, there's a lot of talk nowadays about 'premortem' rather than postmortem," he added.

"Premortem? What is that?" Shyam asked, curious.

Viswa explained, "A premortem is a process where all stakeholders come together and assume that a project might fail or face a disaster. Then, they dissect each critical point in its operational flow to figure out what steps should be taken to overcome the challenges ahead. They also come up with alternate strategies to enhance preparation. In short, it's about seeking a 100 per cent safety and security for the process flow."

"Perfect," said Ankita.

The riverbed was neat and clean. Muni laid out a large bedspread for us to sit on and enjoy some snacks.

Viswa looked at Ankita. "You asked for planning, and that too, pragmatic planning. Isn't it?"

She nodded. "Yes."

She then said, "Planning is what Muni has done. Nobody told him what we might need when we came here by the riverside. He packed this bedspread, snacks, a flask of tea, paper plates, a bottle of water and tissue paper. That's planning."

"Yes, in planning, the mind works forward. It foresees and envisions events, identifies potential needs, articulates the plan and collects resources," Viswa said.

"I'm curious about the river. Can someone tell me if it has any significance?" I asked.

"Well, Amar, that's your job—you're the expert in geography," Viswa said with a smile.

Amar smiled back and began, "Sir, this river is a tributary of the Thamiraparani River, which flows through this part of the state. As you know, the Thamiraparani starts from the Pothigai Hills in the Western Ghats, and it's believed that there are countless herbs along its course. The river is said to be rich in minerals and medicinal herbs. This tributary passes through four villages before reaching here. Except during the two summer months, when the flow reduces, the river is perennial. It provides much-needed portable and irrigation water to these villages. It continues through three more villages before flowing into the main Thamiraparani River."

I was impressed by Amar's in-depth knowledge.

"As such, the villagers have established three major inlet points from this tributary to ensure a steady flow of water for irrigation. We haven't had any conflicts over water management. Now and then, power struggles between people from different areas surface, but it's not due to a shortage of water," Amar added with a smile. "Water has always been a geopolitical issue all over the world, and no place is free from it—not even some islands."

He laughed and continued, "Rajendra Singh, the 'Waterman of India', has predicted that if there is another world war, it will likely be over water."

Viswa said, "At the state, national and global levels, many have used water as a central issue to push political agendas and gain leadership. By claiming control over water sources in their regions, they've created problems that could have been resolved easily. Water is an emotional issue, and in many cases, uninformed farmers have been exploited for political gain."

Everyone nodded in agreement.

"Sir, what are the basic differences between personal, corporate and community leadership?" Ankita asked, sparking a debate.

Viswa responded promptly, "What differences do you notice between living at home, in another state, or in a different country?"

Everyone fell silent, reflecting on the question.

"Your understanding grows as you adapt to others' needs, lifestyles and cultures of others. You begin to realign your personal priorities to accommodate those around you. On a larger scale, this expanded awareness, of knowledge, emotions and behaviour, helps reduce conflicts. The same logic applies across all systems," Viswa said.

"Leadership roles become more passive, tolerant and inclusive when you lead a company, an industry or a larger community," he continued.

"Sir, do you think leaders of larger systems need to be more diplomatic?" Shyam asked.

"Yes. As the operating universe expands and the elements of inclusivity grow, a leader must address the essential human emotional and psychological needs arising from existential and identity crises within these systems. That indeed demands a lot of diplomacy," Viswa said.

"I think someone who practises diplomacy at home could easily become the foreign relations minister of the household," Shyam joked.

We all laughed.

> **REFLECT**
>
> 1. A good leader touches not only the lives of their team members but also those around them.
> 2. Many problems can be solved through collaborative thinking.
> 3. Leaders should develop the ability to examine issues within the ecosystem they operate in.
> 4. Leaders need to understand the basics of diplomacy as they engage with a broader functional landscape.
> 5. Conflict resolution and management require a deeper understanding of the emotions involved.

YOUR LEADERSHIP QUESTIONS

1. What are the three key elements that help build and sustain relationships? Why is relationship management essential for a leader?

2. Lateral thinking enables the creation of alternative pathways to solve problems. Choose any two issues and explain how lateral thinking could help resolve them.

3. Why does understanding the ecosystem in which a problem exists help in finding better solutions?

4. Identify some factors that contribute to conflict and explain how you would approach resolving them as a leader.

5. What are the challenges in fostering collaborative teamwork in any organisation or system? How can these challenges be effectively managed?

13
Farmers and Leaders: Architects of Hope

*In every walk with nature one receives
far more than he seeks.*
—John Muir
Naturalist and author

The next morning, right after breakfast, we decided to visit Viswa's farms to assess the damage from the thunderstorm.

"Will Muni join us at the farm?" I asked.

"Yes," Viswa replied. "He has his own morning routine. At dawn, he walks about 5 km, then feeds the birds and animals. After bathing, he visits the nearby Shiva temple to help the priest clean the premises and spends about 30 minutes meditating there. He hasn't deviated from this routine in years. He doesn't drink tea or coffee; just three glasses of water early in the morning. Before heading to work, he eats two bowls of porridge, without sugar."

I admired Muni's disciplined way of life.

"Sir," Ankita commented, "It is said that two perfect ways to live are through discipline and a positive attitude. He has both.

That's what makes him a self-disciplined, self-motivated leader in this village."

As we walked along the mud roads, I noticed the stagnant water hadn't fully drained. Still, it looked better than many cities after a rain.

"Unfortunately, sir," Amar observed, "we don't have a well-defined, professional health management system in our cities. We're more like firefighters reacting to problems rather than preventing them."

"No, Amar, I disagree," Shyam said. "There is a hierarchy in health management in villages and towns. It may be inadequate, but the real issue is poor execution and management of available resources. Plus, there's a lack of advanced training for professionals and paraprofessionals to deal with these issues, both at the cognitive and operational levels."

"Is this due to poor leadership?" Amar asked.

Viswa replied, "To some extent, yes. There's always a gap between efficiency and effectiveness, between work and skill. The workforce needs to be more compassionate and empathetic. This lethargy exists not only in health but also in education, tourism, hospitality management and other sectors."

"Is it also because of a lack of accountability at multiple leadership levels?" I questioned.

"To some extent, yes," Viswa said. "Workers often feel unaccountable due to job security, believing they won't be challenged. Also, in a populous country like India, one of the main challenges is that the operational burden per worker is high. Some leaders have sincere intentions but lack proper empowerment and processes, while others are active but have questionable motives."

I added, "It's as simple as how I feel when I see something that needs to be cleaned or cleared – 'That is not my job; someone else is paid for it'. And we continue to live with that feeling for too long. We never think that the entire society belongs to us – the sense of ownership is missing."

"I think technology could help handle this better," Viswa said. "India is stepping into its digital operating universe, and this means that problems are getting solved one after the other. Finally, it all depends on the service mentality of people in the field. As the proverb goes, 'Your attitude, not your aptitude, will determine your altitude'."

"Yes," I added, "'Digital India' could open the gateways to better monitoring, mentoring and accountability."

"Correct, leaders should focus on developing the right attitudes alongside empowering them with technology," said Amar.

We walked in silence for a while before Viswa suddenly smiled.

"What made you smile, sir?" asked Ankita.

He pointed to a scarecrow set up in a field. "Do you know what it stands for?"

"Of course, it's meant to scare away birds and insects," she replied.

"Sometimes, we need scarecrows in organisations to drive away lethargy and negativity," he said.

We all chuckled.

"Viswa, aren't the rules and regulations in organisations like scarecrows for the employees?" I asked.

"Not always, and not all of them," he remarked.

"Why did farmers adopt the image of scarecrows?" Shyam asked.

"In farmlands, birds and insects can be a threat to growing crops in an open field. But today, farm owners and managers are turning to high-tech devices with motion sensors to drive away birds from damaging crops," Viswa explained.

"Scarecrows across the country seem to have a similar appearance," commented Amar.

"There are some interesting stories related to them. Ancient Egyptians hung tunics on reeds to scare quails away from the crops. In ancient Greece, wooden statues of Priapus, a god considered a protector of livestock, vegetables and fruits, were placed in crop fields. Japanese farmers even constructed the image of Keiko, the killer whale, to keep sparrows from eating their rice. The design of scarecrows varies depending on local cultures and belief systems," Viswa shared.

"As for organisations, what competence is required to be a scarecrow?" asked Amar.

"No competence is required to be a scarecrow! All you need is to be willing to get branded as one, sometimes at the cost of being ridiculed or considered a fool. Interestingly, one must pity the scarecrow. It's said that the scarecrow is archetypical of the Detached Guardian, someone who protects but ultimately cannot enjoy the fruits of their labour, as it cannot eat the food it guards," Viswa replied.

"In every organisation, there are some scarecrows who disturb the sleep of leaders and followers. They seem to terrorise the leadership," I observed.

Viswa nodded, "That reminds me of a satirical poem on the scarecrow written by the great Lebanese philosopher-poet Kahlil Gibran. It reads:

Once I said to a scarecrow, "You must be tired of standing in this lonely field."

And he said, "The joy of scaring is a deep and lasting one, and I never tire of it."

Said I, after a minute of thought, "It is true; for I too have known that joy."

Said he, "Only those who are stuffed with straw can know it."

Then I left him, not knowing whether he had complimented or belittled me.

A year passed, during which the scarecrow turned philosopher.

And when I passed by him again I saw two crows building a nest under his hat.

"Interesting indeed. But how do we get out of the terrorizing effect they have?" asked Ankita.

"A worthwhile leader should simplify both actions and communication," Viswa answered.

"Absolutely right, Viswa. Very often, leaders who do commendable actions fail to communicate properly and adequately. As a result, the message either doesn't reach the intended audience or doesn't come across in the way it should, missing its purpose," I said.

"A leader needs to be an effective communicator, don't they? Ankita asked.

"Sir, what do you think are the important elements of any meaningful communication?" Shyam asked Viswa.

"You can't give a generalised answer because the goals and objectives of communication vary depending on the place and occasion. Cultural sensitivity is crucial in communication, both on a personal and group level," Viswa said.

He continued, "However, when it comes to leadership communication, three important factors must be kept in mind—focus, clarity and direction. The receiver should clearly understand the message, know the direction to proceed and understand what the bull's-eye is."

"Recently, a lot of studies have been done on NLP—Neuro-linguistic Programming. Leaders, certainly, need to understand the essence of NLP. Words are not loners. Words carry meaning. The relationship between the word and the meaning is like that of the body and soul," I said.

"The word is the matter, while the meaning is the energy," said Amar.

"Words are sources of power and energy, and they can make or break people," said Ankita. "So, an effective leader should be aware of the power of their words and use them appropriately. So, Viswa sir, if a scarecrow is considered a leader, what message does it pass on?"

Everyone smiled.

"What a thought!" Amar commented.

Viswa didn't hesitate to answer. He said, "The message you get is: 'Like a scarecrow, let your purpose cast a protective shadow over the lives of others'."

"Indeed, it is positive thinking," Ankita commented.

Muni joined us and greeted everyone with a *vanakkam*.

"So, Muni, is everything fine?" Viswa asked him.

"Yes, sir, by God's grace, everything is fine. There are only minor issues here and there after the storm, but nothing we can't handle," Muni replied.

I was surprised by Muni's outlook and positive thinking.

"That's the spirit. One should not go on complaining, finding fault or seeing negativity. There are many in leadership positions who see what they don't have rather than what they have to succeed. One gets frustrated with such leaders," Viswa said.

"I fully agree, Viswa. I've seen several people who have a hundred excuses to justify why they failed or couldn't carry out a task, rather than one single reason why they should or could have done it. There is an excellent saying attributed to the playwright George Bernard Shaw, which articulates the fact that 'most people who fail complain that they are the victims of circumstances. Those who get on in this world, are those who go out and look for the right circumstances. And if they can't find them, they make their own'," I said.

"A very powerful statement by Bernard Shaw," said Ankita. Amar and Shyam agreed.

"Optimism is the essence of good leadership," said Amar. "Helen Keller, the American author who struggled through her early life with blindness, remarked, 'Optimism is the faith that leads to achievement. Nothing can be done without hope and confidence'."

"But sir, do you think mere optimism without effort would help?" he asked.

"That doesn't work. That is daydreaming. Optimism must be based on a deep insight into functional dynamics. Otherwise, it will be a castle in the air," I responded.

Muni broke his silence. "Viswa sir, should I organise spraying insecticides today?"

"Muni, last time I told you to ask those sprayers from the nearby town to do this," Viswa replied.

"Sir, this time we will do it manually as usual. The people are ready, and the job will be done," said Muni.

"Sorry, Muni. Even if we wait for one more day, let's telephone them and get those sprayers," Viswa said firmly.

I asked, "Viswa, why are you so inflexible? You could have accepted Muni's suggestion this time."

Viswa glanced at me and the others before speaking, "Folks, remember this: we often become stuck in one habit, one way of thinking, or one method of working. We hesitate to break free from it. Think about the strength of an elephant. But when it is tied to a pillar with a small rope, it feels it can't move, even though it has the power to break open in an instant. This is called 'learned helplessness'. Many of us are trapped by this and refuse to change due to the conditioning of our minds. I wanted to break free from that. By doing so, it will also send a message to Muni."

"'Learned helplessness', what an example! Sir, do you think this 'learned helplessness' will hinder the growth of a leader?" Shyam asked.

"Certainly. It keeps the leader confined to a comfort zone. It nurtures latent fears and gives a sense of false security. It prevents further growth and learning. As Martin Seligman, known as the Father of Positive Psychology, puts it, 'Learned helplessness is the giving-up reaction, the quitting response that follows from the belief that whatever you do doesn't matter'," Viswa replied.

"I agree, sir. Further learning is critical to growth," said Shyam.

"Not only that, Shyam. It should come after unlearning. Unlearning is a very important factor in our growth. If you don't unlearn, you'll face a psychological barrier when it comes to future learning," Viswa replied.

Ankita interrupted, "I just read a line by a great poet, Jack Gilbert: 'We must unlearn the constellations to see the stars'."

"Wow! What a thought! Wonderful to hear," said Amar.

"You all must read a book I admire, *Future Shock* by Alvin Toffler. He writes, 'The illiterate of the 21st century will not be those who cannot read and write, but those who cannot learn, unlearn, and relearn'," Viswa said.

"Is this true and applicable to all fields?" Amar asked.

"Certainly, with no exceptions. We are living in an ever-evolving world where both the convergence and divergence in knowledge are creating more possibilities, more opportunities, and even threats. So, coping with the future is a challenge, where we need a lot of flexibility, effort and energy to cope," Viswa explained.

"Sir, you mentioned convergence and divergence of knowledge. Does it impact agriculture and farming too?" Amar asked.

"No doubt. The future of agriculture is not only more exciting but also more challenging and productive, all of which could be the outcome of innovations in the field," Viswa responded.

"Wow! We need to talk about that too," Amar said.

"Time to go for lunch," Muni reminded us.

REFLECT

1. A leader's attitude determines the altitude they can reach.

2. Scarecrow leadership is fragile and cannot withstand the test of time.

3. Leaders often draw energy from their optimism.

4. "Learned helplessness" is a major obstacle to the future of leadership.

5. Effective communication is one significant requisite for strong leadership.

YOUR LEADERSHIP QUESTIONS

1. If, as a leader, you find that the circumstances for institution-building are inadequate or unsupportive, what steps would you take to create favourable conditions for it?

2. "Optimism is as sweet as honey, but an overdose of it has a retrograde effect." As a leader, what steps would you take to balance optimism with pragmatism to achieve credible results?

3. Identify three challenges you face due to "learned helplessness". How would you overcome them?

4. Examine how the convergence and divergence of knowledge and skills in your field are affecting the way you work. Identify three strategies you would adopt to overcome these challenges.

5. "A leader is an effective communicator." What are two major communication challenges you've observed in yourself, and what steps will you take to address them?

14
Resolving Conflicts in Physical and Human Nature

*Don't judge each day by the harvest you reap
but by the seeds that you plant.*
—**Robert Louis Stevenson**
Novelist and essayist

"Sir, we have the *panchayat* meeting this morning," Muni informed Viswa. "We need to leave in about half an hour."

"We?" Viswa questioned. "What do I have to do there? You're free to go."

"Everyone is expecting you, especially the *panchayat* president, who asked me to personally invite you," Muni replied.

"Is there something important?" Viswa asked.

"Sir, are you aware of the huge conflict between Nataraj and Shankar? The issue is regarding the water flow from Shankar's land to Nataraj's. Shankar has blocked the flow, and it needs to be solved. They're requesting your assistance because neither Nataraj nor Shankar will listen to anyone but you," Muni said.

"That's concerning. They should follow the decision of the *panchayat*. We are all bound by the institutions and systems, not by any individual. The *panchayat* represents the collective wisdom of the community," Viswa said.

"But the *panchayat* leaders hold you in high regard and see you as a guiding light for all their challenges," Muni replied.

Viswa agreed and went with Muni. As the *panchayat* assembled, all the leaders stood for a minute of silent prayer. The *panchayat* president briefly addressed the conflict between Nataraj and Shankar.

Nataraj argued, "It's illegal to block the water flow to my land. Water is the property of the village, and no one should have the right to manipulate or stop its flow. My crops are drying out, and I'll lose everything if I don't get the water."

Shankar retorted, "Yes, water is a common property, while it's flowing on my land, I have the right to block its flow."

"Sorry, but water is a common property wherever it flows," Nataraj replied.

"And gentlemen," Shankar added, Nataraj has blocked the single-track path connecting my land to the main road for the last six months. He claims I can't walk through his land as it amounts to trespassing."

As Viswa listened quietly, Muni kept an eye on his reactions.

The *panchayat* president pleaded, "Viswa sir, you've been our guide and have helped us to resolve so many conflicts before."

Viswa smiled and said, "Can I have an exclusive conversation with both of them outside this assembly for a few minutes?"

"Why not discuss it here in the open assembly?" one member asked.

"If it can resolve the issue, there's no problem with having a private discussion," the president replied.

Both Nataraj and Shankar followed Viswa to a nearby room in the *panchayat* building. The assembly waited patiently, gossiping as they were gone. After 15 minutes, all three returned, smiling.

Viswa said, "Well, the matter has been resolved. Nataraj will allow Shankar and his family members to use the single-track path to the road from his land, and Shankar will unblock the water flow tomorrow. If they have any issues in the future, they will bring it to the *panchayat's* attention before taking any individual action."

The entire assembly clapped, and the *panchayat* members thanked Viswa with folded hands. Nataraj and Shankar both touched Viswa's feet.

"Oh, no," Viswa said, stepping back.

"Sir, you're like an elder brother—senior in both age and wisdom. Please allow us to show our respect," said Shankar.

Muni, who had been silently observing, heaved a sigh of relief. He recalled what Ankita had once said, "Viswa sir is a leader for all reasons and all seasons."

Back home, Muni briefed the others about the entire episode. Viswa walked in and asked, "Is this a replay of what happened?" All of them smiled.

I asked, "Viswa, I'm curious to know how you solved the issue. What was the magic?"

Shyam added, "We want to know the kind of skill you used as a leader."

Viswa smiled and said, "Nothing special. Every conflict is unique. One must examine the issue from its roots and not merely from the surface level. It could be a personal ego clash, irrational resistance to displaying power and authority, or it could stem from an ideological or political stance. In this case, it was essentially an ego problem. The single-track path was being used not only by Shankar's family members but also by his friends and strangers visiting his house. This caused some inconvenience to Nataraj, and his request to make minimal use of the pathway didn't sit well with Shankar. Nataraj then blocked the path, and Shankar retaliated by blocking the water flow to Nataraj's field."

"Sir, how did you bring them to a compromise?" Shyam asked.

"It's simple," Viswa replied. "I spoke to both of them individually for a couple of minutes and praised them for their social skills and compassionate outlook. I helped them understand that they were both right from their own perspectives, but they needed to see the issue from a broader viewpoint. The ability to forgive each other and seek a compromise would relieve them from the animosity and pain they would otherwise carry in the days ahead. Both agreed that it was a trivial issue and were willing to resolve it, provided it was based on mutual respect. I believe that for leaders, negotiation skills are essential. In such mediations, it's important to avoid blame games, as they only hinder negotiations and cause them to stall. One key aspect of conflict resolution is to keep moving forward."

Everyone admired the approach Viswa had taken. He continued. "Every system, every organisation, every community and every country is a forum for multiple conflicts. It's important to bring the concerned persons to a platform for discussion. Sun Tzu, the author of the book *The Art of War* writes, 'Every battle is won before it's ever fought'. A prudent leader understands that."

"Viswa, don't you think that the time and energy a leader spends resolving issues one after the other robs them of their positive energy as well as productive time?" I asked.

Viswa responded, "Of course, that's true. An organisation as well as its leaders should establish a mechanism to resolve conflicts at every stage. First, we need to understand that the emergence and continuation of ego conflicts in an organisation usually signal a lack of role clarity and the boundaries of those roles. While conflicts are not entirely unavoidable, having standard operating procedures and key performance areas for individuals in vital roles is essential. These tools can help identify where things stand and where they should be. I once read a quote by Cullen Hightower, a well-known quotation and quip writer from the United States, 'There's too much said for the sake of argument and too little said for the sake of agreement'. A great leader should know how to differentiate between the two and handle them accordingly."

"Sir, do you think conflicts are by-products of invisible competition among individuals or systems?" Shyam asked.

"It could be, or it may not be. If conflicts arise from visible or invisible competition, you need to get to the root of the problem to set things right. Often, competitions stems from unclear goals or from attempting to impose one's ideals, goals or achievements onto others. It may also result from trying to emulate the success or appeal of other people or brands," Viswa said.

"Well said," commented Amar.

"Sir, is there a well-grounded approach to resolving conflicts?" asked Ankita.

Viswa smiled and said, "You're asking for a magic wand. As long as humans exist, there will be conflicts too. Nature has taught us many ways to handle them. For example, thorns and roses exist on the same plant, a gentle breeze can turn into a tornado and the same land that creates can also destroy through earthquakes or other disasters. Conflicts are present in nature and wildlife as well. But Martin Luther King Jr, the American activist for equal rights through non-violent protests, offered a meaningful response to your question. He says, '…man must evolve for all human conflict a method which rejects revenge, aggression and retaliation. The foundation of such a method is love'. The world offers immense opportunities for everyone, and once you realise this, you become your own competitor. That's the best thing that could happen."

"Is it wise to compete with oneself, or should one avoid it? Doesn't it create stress and tension? Doesn't it always put one in a poor light?" asked Amar.

"At the leadership level, it is essential to compete with one's own self," said Viswa. "As Jay-Z, the American rapper, aptly says, 'I look in the mirror, my only opponent'."

Muni interrupted the discussion and told Viswa, "Sir, Rosiah, accompanied by two other people, has come to see you."

"What do they want?" Viswa asked.

Although Muni knew the reason for their visit, he pretended ignorance. Viswa excused himself and went out to meet them.

After exchanging pleasantries, Rosiah said, "We farmers are facing a severe problem. As you know, I'm a wholesale trader of vegetables. The price of tomatoes has fallen drastically—the market price is now just Rs 2 per kilogramme. The wholesale cost is still low, and there are no buyers. Since tomatoes will decay if held too long, I thought I must take your advice."

"Have you explored other city markets?" Viswa asked.

"Sir, even if I include the shipping cost, I will still suffer a loss," Rosiah replied.

"Did you talk to the local agricultural marketing officer?" Viswa asked.

"Well, sir, they have no suggestions. They just play into the hands of the agents," Rosiah said.

"Alright. Hold on for a few minutes. Muni, can you call Shyam?" Viswa asked.

As Shyam walked in, Viswa said, "Sit down, Shyam. Rosiah has a genuine problem with tomatoes. The market prices have dropped drastically, and he's facing a loss. You could do two things. First, reach out to the cold storage facility in the city and find out if they can store the tomatoes for about 15 days. By then, I hope the prices will recover. Second, write a petition to the collector on Rosiah's behalf, highlighting the issue and seeking an agricultural subsidy due to the price drop. It will be a great learning experience for you. More importantly, conduct some research to understand why the price dropped suddenly and when it's likely to revert."

Rosiah smiled and said gratefully, "That's why we came to you. We knew you would find a way out."

I was watching Viswa from a distance, and was overjoyed that I was his friend.

"There's another problem too, sir," Rosiah continued, pointing to one of the men with him. "He is Santhanu, my friend. He grows jasmine flowers, among other crops. He heard from someone that there's a quality branding initiative for jasmine in Madurai, and there's a huge international market for these flowers. Of course, they only bloom in certain seasons, but he's interested in exporting them, if possible. If you could advise us on this, it would be a great favour."

"Oh, yes. Jasmine flowers from certain areas of Madurai are ISO (International Organisation for Standardisation) certified and are exported. I'm glad to hear you want to focus on growing quality products," Viswa said. "Muni, this is Amar's turn. Please call him."

When Amar arrived, Viswa said, "Amar, go to the ISO website and find out the criteria for ISO certification for jasmine flowers. Also, try to gather information on the process of standardising agricultural products. Pass that on to Santhanu. You can also email the state's agriculture department to learn about its role in the production and export of jasmine flowers. You must also visit the website of the Export Promotion Council of India for more details. Work on this. Of course, Ankita can also help you."

The visitors left, satisfied and grateful.

I remarked, "I'm amazed with the way you delegated the work to your team in a split second."

"Remember, Bala," Viswa said, "you should never chase a black hare in a wild forest at night. A leader who tries to do so will certainly fail. And a leader won't have enough time for everything if they really want to grow."

"You're right Viswa. I just wanted to hear your views on how a leader should manage their time," I said.

"Good question. First, remember this: time doesn't manage us—we manage time. I came across the four D principles of time management in one of John C. Maxwell's books: Dump, Delay, Delegate, Do. If you look at the work on your desk, you'll find that one-fourth of it isn't worth doing at all—just dump it.

Now, another one-fourth of the remaining work doesn't need immediate attention—it can wait. That leaves you with 50 per cent of the work. You'll find that you don't need to do everything. Half of what's left can be delegated to others. Finally, you're left with 25 per cent of the total work. Just do it.

"Wow!" I said, jumping up from my seat. Muni stood there as if he fully understood the essence of what Viswa had said.

All three youngsters were busy taking notes on their mobiles.

> **REFLECT**
>
> 1. Conflict management and resolution are essential leadership skills.
> 2. Resolving conflict requires a deep understanding of the issue and an empathetic approach.
> 3. Competing with oneself fosters personal and professional growth.
> 4. An effective leader delegates responsibilities to help team members develop their own leadership skills.
> 5. A good leader understands the value of using time wisely for productive outcomes.

YOUR LEADERSHIP QUESTIONS

1. Why do you think conflicts are a natural part of a vibrant system? Suggest two key conflict management skills.

2. How does empathy play a significant role in conflict resolution? Illustrate with an example.

3. How does setting priorities contribute to effective time management?

4. What are common challenges in delegating work to team members, and how can these be prevented?

5. Do our village products have potential for export? Suggest two products and explain why they are suitable.

15
The Glory of Agricultural Leadership

Good leaders build products. Great leaders build cultures.
Good leaders deliver results. Great leaders develop people.
Good leaders have vision. Great leaders have values.
Good leaders are role models at work. Great leaders
are role models in life.

—Adam Grant
American author

As we all sat down to continue our discussion, the siren of an ambulance interrupted us.

"Muni, why is an ambulance coming here? Can you find out?" Viswa asked.

Muni rushed outside and returned a few minutes later.

"Sir, Velu was bitten by a snake. The *panchayat* president called for the ambulance to take him to the city hospital," he said.

"Have they given him any first aid?" Viswa asked.

"Yes, sir. The people who were present gave him an extract of a herb they say is an antidote for snake venom. They also tied a string around the area where he was bitten," Muni responded.

"Viswa, do people here have any expertise in first aid practices?" I asked.

"Yes. They've been dealing with these problems for ages and have several local practices to save people before they're rushed to the hospital," Muni replied.

"Was it a poisonous snake?" Viswa asked.

"Yes, it was. It was a *kannadi viriyan*," said Muni.

"Oh, a Russell's Viper or *Daboia russelii*," Ankita noted.

"Let's see if we can be of any help," Viswa said, and we all followed him and Muni.

The ambulance had arrived, and Velu was being helped into it. He was crying out in excruciating pain, and his wife followed him to the ambulance.

"Ponni," Viswa addressed Velu's wife. "Do you have money, or do you want me to accompany you?"

"Sir, I've Rs 200 with me. I'll arrange for some more money on the way to the hospital. Both my sister and brother-in-law are coming with me," Ponni replied.

Viswa took out Rs 2,000 from his wallet. "Take this for any emergency. Don't waste your time on the way." Ponni was moved, and the *panchayat* president smiled at Viswa, thanking him.

We returned home. "These are the challenges of rural life—insect bites, allergies and the absence of proper or quick medical attention," Amar lamented.

"Can't the local leaders do something about it?" I asked.

"All that you need is not necessarily a community leader because, oftentimes, it is tainted with political colour. We need compassionate leadership," Viswa said.

"Can't community leadership also be compassionate?" Ankita intervened.

"Of course, it can. But the convergence is rare, although community leadership with compassionate leadership is an ideal combination," Viswa said.

"Sir, can you briefly tell us what compassion is?" Amar asked.

Viswa commented, "That's a good question. Compassion is not pity, nor is it sympathy. It is empathy in action. As psychologist and author Daniel Goleman put it, 'True compassion means not only feeling another's pain but also being moved to help relieve it'."

The conversation continued over the coffee table.

"Viswa, you mentioned the book *The Future Shock* by Alvin Toffler. Has he said anything about the future of agriculture and farming?" I asked, curious.

"Yes, he has," Viswa replied. "But before I get to Toffler, let me share what M.S. Swaminathan once said: 'If agriculture goes wrong, nothing else will have a chance to go right'. That highlights just how critical agriculture is."

"Are there enough innovations in agriculture today to open doors for new leadership models?" I asked.

"Bala, rather than answer that myself," Viswa said with a smile, "I'd prefer to let the young agricultural entrepreneurs respond to that."

Shyam began, "Sir, your question could invite multiple answers, which might take hours even if we kept it brief. I'd say innovations in agriculture are abundant. As a result, leadership across agricultural production, management, marketing, and processing is likely to undergo key reforms. These changes will be driven not only by traditional methods but increasingly by digital technologies. From seeding to harvesting, and of course in marketing, everything is moving toward digitalisation. The future of agriculture will be guided more by science and logic than by past knowledge and experience."

Ankita said reflectively, "That sounds great."

Amar added, "Sir, for example, seed banks now control seed quality and productivity. Mobile soil-testing labs provide instant results. There are fertilizer combinations tailored to specific soils and crops, affordable organic farming methods and weather forecasts updated hourly. Modern harvesting techniques require minimal labour, and advanced storage methods prevent waste and losses from rodents and other threats. Warning systems track insect movements and hazards, while national market indicators help shift products to where demand and prices favour farmers. I think our perception of agriculture is changing."

Shyam explained, "Sir, many innovations in processing have been tipping points for transformation in agricultural business and leadership. For example, horticulture, floriculture, pisciculture, aquaculture and the like have generated millions of dollars through exports. Food processing, packaging and related industries have opened new gateways for entrepreneurship and fresh ideas ."

Viswa interjected, "Floriculture has huge export potential."

Shyam continued, "There are even more innovations underway. Precision agriculture uses data collection and analysis to help farmers optimise soil quality and productivity. Source-use efficiency focuses on sustainability, profitability, productivity and quality. Aeroponics suspends roots in air, with emitters periodically spraying them with water and nutrients. Genetic modifications have also played an interesting role in developing new products."

"Just hold on," I said. "There have been many concerns about GM (genetically modified) food and seeds. Am I right?"

"You're right, sir. I think most of those concerns have been settled now. Changes and transformations always bring fears, apprehensions and counterattacks. But largely, many of these apprehensions have been cleared by international agencies," Ankita responded.

Amar interrupted, "Sir, there's another innovation. It's called indoor vertical farming. In this facility, farm produce is grown by stacking crops above each other in a closed and controlled environment. The technology uses growing shelves mounted vertically to increase crop yields in limited spaces."

"Enough I suppose," I said, and everyone laughed.

Viswa said, "Bala, Thomas Jefferson, the third President of the United States, once said: 'Agriculture is our wisest pursuit because it will, in the end, contribute most to real wealth, good morals, and happiness.'"

Ankita added, "You're right, sir. 'No race can prosper until it learns there is as much dignity in tilling a field as in writing a poem,' said Booker T. Washington, the American author and educator."

I asked, "So, Viswa, can we say that agricultural leadership is a forerunner of all other types of leadership?"

Viswa quickly responded, "No, I wouldn't like to stir a controversy by setting debatable priorities. It may largely be true. There's a lot of weight in that statement because agriculture is one of the most primitive, yet premier, professions that we can practice."

He continued, "I can certainly say that agricultural leadership encompasses several types of leadership, like leading from the front, from the back, from the side, and leading by the leaders' presence or absence."

Amar added, "Sir, it includes compassionate and ethical leadership."

"Well said," Shyam observed.

Muni walked in with a bowl of cut-up fruit. He smiled and said, "Velu is out of danger now. The doctors administered some injections and treated him. They also praised the timely action of bringing him to the hospital."

We all heaved sighs of relief.

"It appears that farming is the most glorious work. It gives full satisfaction, plenty of personalised and collaborative learning, teamwork, participative leadership and many other benefits," I said.

Then Viswa gave a long lecture, mostly directed at me, as he became quite emotional about what he was doing.

He said, "You know, Bala, what George Washington, the first President of the United States, said? 'I would rather be on my farm than be emperor of the world'. Farming isn't a job that just anyone can do. It's a specialised skill that calls for a congruence of knowledge, talents and aptitudes. A farmer must integrate thinking, analysis and emotions, along with social, marketing and entrepreneurial skills, to succeed and become a complete farmer. Patience, perseverance, and precision are crucial in executing the work.

"A farmer also needs a deep understanding of the ecosystem and how it responds to human efforts. He has to understand how the plant and animal kingdoms collaborate with him in the effective discharge of his intended work. And to be honest, many of these insights can't be fully taught—they come through experience. Of course, experience and academic research can trigger this passion and remove the roadblocks along the way."

"Can I add something to what you said, sir?" Shyam asked politely.

"Of course. Knowledge is always meant to be shared. A person who hoards knowledge without sharing it can never grow to become an influential leader," Viswa said.

Shyam continued, "A farmer must be conscious of the safety of the produce and sensitive to the effects of both organic and inorganic interventions on plant growth. He should understand the life cycle of the crops and be aware of how external factors, such as floods, droughts, unpredictable weather and insect infestations, impact their development. These challenges are constant at every stage, but they must be faced with courage and conviction. Losses can be minimised if proper waste management techniques are adopted. Therefore, the farmer must be wise, compassionate, smart, sensitive and authentic in his relationship with the farmland. From seeding to harvesting, he must exhibit a wide range of skills, whether learnt through formal education or informal methods."

"That perfectly captures the essence of leadership in farming," I responded.

> **REFLECT**
>
> 1. Compassion is an essential quality for a humanist leader.
>
> 2. Reaching out to people enhances social and emotional cohesion.
>
> 3. Leaders must understand that actions speak louder than words.
>
> 4. A leader shares knowledge with the entire team so everyone can benefit.
>
> 5. A leader should develop a deep understanding of the ecosystem in which they operate.

YOUR LEADERSHIP QUESTIONS

1. Identify five important leadership skills demonstrated by a farmer in this chapter.

2. Why is agricultural leadership said to encompass various types of leadership?

3. Why is farming not a job for just anyone? Share two or three examples from your personal experience.

4. Name two recent technological interventions that have impacted agriculture.

5. Provide one example of experiential learning in leadership through farming.

16
Farmers and Leaders Celebrate Diversity

Most of the important things in the world have been accomplished by people who have kept on trying when there seemed to be no hope at all.
—Dale Carnegie
Author and professional trainer in leadership

It was 5 p.m. I was sitting in the small hall, reading some of the magazines on the table. A couple of them were related to agricultural marketing and food processing.

Amar, who was also in the hall with Ankita, told her, "Get ready in the next 10 minutes. We'll have to leave."

"Okay, I'll be ready," Ankita replied.

I looked up in surprise. Viswa noticed my expression and said, "They have a class at 5.30 p.m. They conduct life skill classes for the local community once a week, on Wednesdays. Tomorrow, I'll give a lesson on financial literacy."

"Wow!" I exclaimed. "But then, what is your objective? Do you think you'll be able to improve them?"

Viswa laughed. "Initially, I doubted their willingness and the kind of cooperation they'd offer. But later, I found there was a huge response. They have an innate desire to be on par with their counterparts in the cities. After all, they haven't had the opportunities people in the cities do. Once they grew, a psychological barrier made it hard for them to view themselves as learners. We had to work hard to break that barrier. Essentially, it's an attempt to empower them so that they're not cheated. They no longer have to stand before others as ignorant people."

"The aim is much higher than just helping them move beyond their own idea of who they are. They feel empowered and celebrate their self-confidence," said Ankita.

"And what do you teach them?" I asked.

"Amar and Ankita teach them basic language skills, communication, digital skills, interpersonal skills, conflict resolution and management skills. To be honest, most of them already know all these, but they don't realise it. Once you ignite the fire in their bellies, the job is done," Viswa said.

"What strategy do you use?" I asked Amar.

"It's simple. I present case studies from their own lives and make them discuss them. They get deeply involved and that gives them a sense of ownership value in their learning."

"Sir," Shyam intervened. "Their ability to learn is far greater than what they are taught. There's a huge opportunity to take them to the next level."

"Interestingly, I've noticed a lot of curiosity among women, especially when it comes to financial literacy. Their eagerness to learn is high. I think they've been kept in the dark for several decades," said Viswa.

"Kept at home, you mean?" Shyam asked.

Ankita replied, "No, just that. Even though they worked at home, the roles they performed were remarkable. They nurtured our culture through various art forms, demonstrated creativity and innovation in everything they did. Unfortunately, their contributions are often undervalued compared to those of professionals working in a 9-to-5 jobs. We've often seen people who spoke foreign languages as the bearers of wisdom. The real issue lies with our mindset."

"I agree with you, Ankita," Viswa chimed in. "Take Muni's wife, for example. She's a talented folklore singer, and people love listening to her. She also uses coconut fibres to make ropes, bags, etc., and coconut shells to create cups, which she decorates with beautiful paintings. She earns from her work, but unfortunately, we don't value what rural women do adequately. It's essential to acknowledge their contributions in informal, unorganized sectors just as much as we do for the formal city professions."

"You've put it so simply, Viswa. But you may not realise the kind of service you're providing to the community—without any expectation, profit or recognition," I said.

Viswa smiled, "You are right, Bala. If work is done only for profit and recognition, it becomes just a job, not leadership."

I understood and agreed with Viswa's perspective.

"And what do you do for financial literacy?" I asked Viswa.

He explained "Money has become essential in our lives, but for many, managing it is a major challenge. Most farmers don't have a regular cash flow. They only earn during certain periods, mostly after harvest. Because of this, they often have to take loans, borrowing from private lenders at high interest rates, and end up getting cheated. Sometimes, they can't repay and lose their investments and assets. Many aren't aware of how banks can help them to raise funds or how they can systematically improve their lives. They could invest their small savings in systematic investment plans (SIPs) to build money for the future. They also need to learn about life insurance and the financial aspects of crop sales. When they have money, they often keep it at home, and sometimes during floods or natural disasters, they lose it. I educate them on all these issues, and they're relieved to find out that the government and various organisations can support and assist them."

He continued, "In a fast-changing world, where the shelf life of knowledge and skills is becoming shorter, transformation is inevitable. If you don't embrace change, you'll be left behind. Prof. Leon C. Megginson expressed this concept well with the following lines: According to Darwin's Origin of Species, it is not the most intellectual of the species that survives; it is not the strongest that survives; but the species that survives is the one that is able best to adapt and adjust to the changing environment in which it finds itself'."

I fully agreed, "There should be no individual who is unable to manage their own finances and wealth."

Viswa said, "And now, with mobile banking, online transfers and digital payments through various gateways, staying relevant in financial transactions is essential."

"Three key aspects of good governance, and thus leadership, are education, empowerment and enforcement," he continued. "An effective leader addresses all these three issues simultaneously and proportionately."

I commented, "You, Viswa, embody a holistic leadership. If even one of these aspects is lacking or disproportionate, it creates problems in the community."

"A good leader doesn't brand themselves as such. It's the followership that defines a leader. Moreover, this followership should not be driven by popularity or like target rating points (TRPs), but should be genuine, dedicated, spiritual and mindful. I fully endorse the views of Ralph Nader, the American activist, who said, 'I start with the premise that the function of leadership is to produce more leaders, not more followers,'" Viswa responded.

I said, "Unfortunately, today, branding has become a massive business. Individuals, institutions and organisations spend a great deal of money to position themselves in a visible world."

Viswa replied, "That may be a compulsive demand for business expansion. But many times, several myths are created around the profiles of individuals and organisations, which are fake, misleading and harmful to society. For example, the laurels, awards and encomiums showered on some people hailed as leaders and pioneers can be annoying...."

Ankita walked in and joined the discussion, saying, "This village has great potential for developing many leaders."

Muni, as always, looked happy and his face lit up whenever people spoke about the capabilities of the villagers. He gestured to Ankita, but I didn't quite understand what he meant.

Seeing my confusion, Ankita explained, "Sir, Muni wants me to tell you about the major change Viswa sir has brought to this village. He personally spoke to different groups to eradicate smoking and tobacco use here, and he was successful in achieving this goal."

Hearing this, I was taken aback. "What?"

Ankita continued, "Yes, sir. He held several meetings to

explain the impact of tobacco on health. Now, no one smokes in this village, and no shop sells tobacco products."

"Now he is advocating against alcohol. He has managed to stop many people from drinking, but it will take some time to make a full impact, as most villagers have been drinking toddy for years. After a long day's work, they seek comfort in alcohol," Shyam said.

"That's not comfort. It's discomfort—for both the self and others," I replied.

I then said, "Don't get annoyed with me if I ask you a difficult question." Viswa smiled.

I continued, "Don't you think these efforts may lead to developing a sheep (herd) mentality in people or even spark a rebellion?"

"Certainly not. If you think that people who are quiet, well-behaved, simple, patient, attentive listeners and have a service mentality are showing sheep mentality you are mistaken. Winning over the self is far more important than winning over others. A good leader first conquers themselves before they conquer others," Viswa responded.

"Can we say that sheep mentality is more of a tendency to fall in line with others' thinking out of fear, laziness, stupidity, insecurity, or a lack of intellect?" asked Ankita.

"There could be other reasons. For example, people are often emotionally swayed by the speech, actions or style of popular yet less credible leaders. As a result, they choose not to question, contemplate or evaluate them," Viswa replied.

"Does institution-building require a community rather than an individual? Is that correct?" Amar asked.

"Not necessarily. Both individuals and communities contribute to institution-building. Sometimes, powerful, committed and passionate individuals add great value to an institution, putting it in a system and giving it a distinct identity. But without the involvement of communities, institutions lose their relevance," Viswa explained.

"What do you see in an institution beyond its infrastructure?" I asked.

"A great question," Viswa said. "An institution is not just about infrastructure. It represents a concept, a culture, a philosophy, a thought architecture and a way of life. It's a place where people celebrate their ideas, the perceptions that arise from them, their values, and engage with a vision of what they aspire to achieve."

"Sir, I can see the value of your statement in the history of several institutions worldwide," said Shyam.

Viswa replied, "Institutional leadership, therefore, seeks to establish an ideated version of individual leadership aimed at the public good, contributing to the betterment of a community or a state."

"What is the typical relationship between individual leadership and institutional leadership?" asked Ankita, adding, "Jean Monnet, the French administrator, said, 'Nothing is possible without men, but nothing lasts without institutions…'."

"An interesting question, Ankita. In my opinion, both are inclusive as well as exclusive. Inclusive, because an individual contributes to, associates with, and participates in the progressive growth of the institution; exclusive, because the individual often outgrows the institutional thought architecture and experiences the freedom of expression in the world beyond the institution," Viswa responded. His insights were engrossing the team.

Muni was listening patiently to the discussion and gestured something to Ankita.

"*Vanakkam* sister. We are on the way," four middle-aged women told Ankita as they made their way to the class. Amar and Ankita went along with them.

Left alone with Viswa, I remarked, "Viswa, don't you find it interesting that such efforts bring people with diverse interests, aptitudes and commitments together on the same platform for a common goal and a common good?"

Viswa replied. "Yes. Unity in diversity is a concept that operates at almost all levels. The role of a leader is to forge that unity within a group."

I asked, "What kind of unity? Conceptual or emotional?"

He laughed. "Both. Strong leaders target both because they complement each other. If you observe, many political platforms focus on emotional unity, which then leads to conceptual unity. On the other hand, many religious platforms emphasize conceptual unity, which leads to emotional unity. This is my perception. There could be other views too."

"Are there any proven case studies to support these views?" I questioned.

"I don't know. My views are based on my understanding. But in general, a good leader is one who moulds a shared vision so that the commitment to that vision creates their following. Interestingly, Martin Luther King, Jr. once remarked, 'A genuine leader is not a searcher for consensus but a moulder of consensus'."

"What is the number of conflicts in villages, and how are they resolved? Is the local leadership adequately empowered to deal with these?" I asked.

Viswa shared his perspective, "Conflicts are a natural part of life, and wherever human beings exist, conflict follows. It's also an aspect of nature that teaches us how to address disagreements or disputes. There are valuable lessons to be learnt from the leadership models the villagers have embraced over the years. One such model is seeking the wisdom of the elderly at home and following their guidance, as they are often the undeclared leaders, believed to be endowed with wisdom. Their impartial approach to problem-solving and conflict resolution is remarkable.

"Another model is that of the *panchayat*, rooted in the concept of *pancha parameshwar*—the five aspects of the Supreme Being. When major issues arise, the *panchayat* comes together, and the majority opinion is binding on the community. This structure not only provides an opportunity to build a consensus on decisions as well as between divided opinions but also respects the majority's decision, which is an authentic exercise in democracy."

He continued, "At times, when the views or decisions of a leader are dismissed by those above them—be it senior figures, boards, management or even the judiciary—the leader must handle the situation with grace. A good leader sets aside personal views when addressing matters that affect the common good of the community."

"Does every leader feel the pressure to comply with family, friends and external influences when the situation calls for it?" I asked.

"Unfortunately, yes. However, a wise leader knows how to distance themselves from such external pressures. A leader should never allow personal interests or outside influences to interfere with their leadership. If they do, it signals the death bell of their tenure."

We concluded the conversation with a mutual commitment to take these insights to the next level.

> **REFLECT**
>
> 1. A leader is accountable not only to themselves but to the community they belong to.
> 2. Institutions represent a concept, a vision and a philosophy, and build a culture around them.
> 3. Institutional leadership is critical to the progress of a community.
> 4. An effective leader is a moulder of consensus and uses collective wisdom to celebrate unity in diversity.
> 5. Leadership is essentially a responsibility, not merely an authority.

YOUR LEADERSHIP QUESTIONS

1. Why do you think self-learning and self-directed learning are important for a leader? How would you guide your team towards these goals?

2. What challenges arise in building consensus within a team? As a leader, what steps would you take to overcome them?

3. What precautions should a leader take when delegating responsibilities and authority?

4. Do you believe celebrating diversity within a team is important? How would you foster and encourage diversity?

5. Where should a leader draw the line to ensure their personal branding does not overshadow institutional branding?

17
Leadership: A Journey of Continuous Learning

*A leader is best when people barely know he exists,
when his work is done, his aim fulfilled,
they will say: we did it ourselves.*
—Lao Tzu
Chinese writer and philosopher

Muni hinted that harvesting had begun in many fields and suggested I may like to visit those farms to see how it was done.

"Indeed, the harvest is a reward for the sweat and labour of the people, so they should be happy," I responded.

"Bala, I must remind you that the quality and quantity of the harvest depend on the efforts invested in various factors like the soil, seeds, manure, weather conditions and, of course, the cultivation process," Viswa explained.

"Cultivation, I think, is really the critical aspect for quality growth," I said.

"Yes," Viswa replied. "The word 'cultivation' applies both to farming as well as human development. As much as the right inputs and a conducive environment are crucial for growing quality crops, so too is the cultivation of the human mind. You can extend this idea to the cultivation of mindsets for prospective leaders."

"How true!" Amar exclaimed.

"What do you think are the nutrients for cultivating the human mind?" I asked.

"Well, there are many," Viswa said. "I'm not going to dwell on them. Nurturing and cultivating the human mind are part of a sacred process. Both family and education play vital roles in it. That's why people strive for a positive, supportive, and happy family and school environment to help cultivate the mind."

Ankita intervened, "Sir, I've been recently studying several articles on neurocognitive research, and they have provided me a lot of information. Some of the findings are indeed thrilling and mind-boggling. Specifically, the research on neuroplasticity, which explains that the brain is essentially plastic and can continuously learn, unlearn and relearn, has opened new ways of understanding cognition, learning and pedagogy."

"You're absolutely right," I said. "I read about the concept of a 'cognitive shift', which explains how the brain can learn through other associated cells, even when the inputs from the sense organs lack connectivity in the brain."

"The fact that learning can be lifelong and re-engineered offers great opportunities for enhancing, empowering and engaging with learning, regardless of the circumstances. For leaders, this is a boon, as leadership is a continuous and sustained learning process. Only those who are committed to learning can lead," I added.

"Sir, what about the cultivation of leadership?" asked Amar.

"Cultivating leadership is both an art and a science. Unique and tailored inputs are required for each individual aspiring to be a leader. It all depends on assessing the personal profile of each individual. As John C. Maxwell, an author known for his work on leadership, says, 'The pessimist complains about the wind. The optimist expects it to change. The leader adjusts the sails'," Viswa said.

Ankita chuckled, "It's like the unique treatment needed for each type of crop."

"Of course. Every plant has its own fingerprint," said Viswa.

"Don't people have natural abilities and talents? Aren't those enough?" Shyam asked.

Viswa responded with an interesting observation by writer John Ruskin, who said, 'Natural abilities can almost compensate for the want of every kind of cultivation, but no cultivation of the mind can make up for the want of natural abilities'.

"What are the significant factors that help in cultivating the mind?" Shyam inquired.

"It is said that the four steps of harvesting are reaping, threshing, cleaning and transporting. Honestly, all these steps are very relevant to the cultivation of the mind for leadership—assimilation of knowledge and skills, the powers of discrimination, clarity, communication and so on…," Viswa explained.

"Do you think leaders can acquire these qualities from formal learning systems?" Amar asked.

"Some of them, yes, but not all," Viswa responded. "Many are gathered through experience. Since leadership is a journey and not a destination, a leader tends to cultivate the mind every day, every moment and through every experience. As William

Godwin, writer and political philosopher, pointed out, 'The extent of our progress in the cultivation of knowledge is unlimited'."

"What does an effective leader harvest?" Amar asked cheekily.

"I'll leave this question to you for suggestions," Viswa smiled.

"Success," said Amar.

"Excellence," said Shyam.

"Productivity," smiled Ankita.

"Wealth," I suggested.

"Second round," said Viswa.

"Happiness," said Amar.

"Shared vision," Ankita remarked.

"Wisdom," Shyam offered.

"Every one of these things," I said.

Viswa laughed. "All of you are right. The harvest brings all these components. Meister Eckhart, the German Catholic priest and mystic, said, 'What we plant in the soil of contemplation, we shall reap in the harvest of action'. But remember, the harvest is the outcome of an event or process. That's not the end of the game. The soil must be ready for the next cultivation and harvest. Therefore, the greatest asset a leader has is a willing, conscious and passionate team who, after the harvest, asks, 'What's next?'"

I quipped, "I think Viswa will not let the team rest and relax."

He responded, "They're built into the schedule because, for a good leader, the joy of work is relaxation in itself."

I replied, "As author Richard Bach says in his book, *Illusions: The Adventures of a Reluctant Messiah*, 'What the caterpillar calls the end of the world, the master calls the butterfly'."

"That puts it in a nutshell," said Viswa.

We all moved towards the field to watch the harvesting operations. As we walked, Viswa remarked, "Folks, you know, in ancient times, elephants were used for post-harvest operations. We have references to this from the history of Chola Empire."

"As such, harvest festivals are celebrated all over the world, in one form or the other," I said.

"Farmers have a unique way of paying homage to the soil, rivers, the sun and the animal kingdom. These celebrations reinforce the idea that we live in an interdependent world, in harmony with nature and Mother Earth," Viswa said.

Ankita was quick in her reply, "That brings us to another important aspect of leadership's responsibility towards Mother Earth and nature—the need to maintain their purity and balance."

"Absolutely," said Amar. "I think that's an aspect we've neglected in our responsibilities and, as a result, we've brought this planet into serious trouble."

"All types of pollution are haunting us—pollution of the land, water and air …" observed Shyam.

"You missed noise pollution—that's also part of the elements that challenge us," Ankita said.

"The unfortunate part of leadership is its focus on business development and profit-making, even at the cost of harming nature. Leaders of every system and organisation owe a moral responsibility to nature and its purity. It is important for them to incorporate this vision, along with strategies and programmes, into their development efforts, so that issues like pollution, deforestation and waste management are part of their design," I said.

"Certainly, sir. Unless these concerns become integral to our thinking, they won't be incorporated into processes and programmes," Amar said.

"Do you suggest that it should be a part of their design thinking?" Shyam asked.

"Yes. Leaders need exposure to the processes and procedures of design thinking," I replied.

"Sir, what's your take on design thinking and why do leaders need it?" a curious Amar asked Viswa.

"Amar, design thinking is a conscious effort to improve the basics of a process to get better results through innovation. To put it more exactly, design thinking is the process of creating a design by introducing innovations," Viswa said.

"Normally, when people talk about design, they refer to the art, style and the way something is represented," said Shyam.

"That's possibly a mistaken notion about design thinking. Richard van der Laken, the graphic designer and founder of the 'What Design Can Do' movement for societal change, says: 'Designers can do more than make things pretty. Design is more than perfume, aesthetics and trends'," Viswa replied.

"How does it consider the needs, expectations, aspirations and mental images of the consumer or practitioner? Is it a top-down process?" asked Amar.

"No Amar, it is not. It's an inclusive, participative process. David M. Kelley, a designer and academic, argues, 'The main tenet of design thinking is empathy for the people you're trying to design for'. Leadership is the same thing—building empathy for the people that you're entrusted to help," Viswa clarified.

"That sounds interesting," said Ankita.

"Yes," continued Viswa, "An automobile designer, Freeman Thomas, has a great response to this question. He says, 'Good design begins with honesty, asks tough questions, comes from collaboration and from trusting your intuition'."

I shared my thoughts, "It's refreshing to hear the word 'intuition'—a mental faculty that has been relied upon for centuries, yet is now being sidelined."

"I agree. Once Albert Einstein, the theoretical physicist known for his theory of relativity, emphasized the power of intuition by saying, 'The only real valuable thing is intuition'," Viswa said.

"How do you see the importance of intuition in leadership?" I asked.

Viswa responded, "Bala, to understand this, you must reflect on the words of none other than Steve Jobs, the co-founder of Apple Inc., and an innovator par excellence who led a revolution in technology. He said, 'You can't connect the dots looking forward; you can only connect them looking backwards. So you have to trust that the dots will somehow connect in your future. You have to trust in something—your gut, destiny, life, karma, whatever. This approach has never let me down, and it has made all the difference in my life.'"

"Is that faith or intuition?" I countered.

"Ha, ha, that's a good question! Have you heard what William Wordsworth, the famous English poet, said? He put it simply: 'Faith is a passionate intuition'," Viswa replied.

"What a fitting reply!" Ankita commented.

"If we talk about faith, we're heading in a different direction—it takes us into the realm of spirituality. Does leadership have anything to do with spirituality?" I asked.

"Bala, it seems you're on fire today with your thought-provoking questions! You must have heard of Deepak Chopra, the renowned international trainer of spirituality, health and healing. More than that, he's the author of several bestselling books translated into multiple languages. He offers this insight on spirituality in leadership: 'Enlightened leadership is spiritual if we understand spirituality not as some kind of religious dogma or ideology but as the domain of awareness where we experience values like truth, goodness, beauty, love and compassion, and intuition, creativity, insight and focused attention'."

I disagreed, "But how does a leader in farming relate to the spiritual domain of leadership? They're just pragmatic, working the land and fighting with the tools at hand."

Viswa observed, "Daniel Webster, a prominent American statesman and orator of the 1800s, said, 'Let us not forget that the cultivation of the earth is the most important labour of man. When tillage begins, other arts will follow. The farmers, therefore, are the founders of civilization.' And as the founders of civilization, farmers are also architects of life processes, which cannot exclude the domain of spirituality."

"Sir, if I may intervene," Ankita said, "I once read the words of British author James Allen, who said 'The law of harvest is to reap more than you sow. Sow an act, and you reap a habit. Sow a habit and you reap a character. Sow a character and you reap a destiny'."

"Good example, Ankita," Viswa acknowledged. "After all, the harvest helps us live safely and securely during difficult times, doesn't it?"

"It's time for us to enjoy a cup of tea," said Shyam.

"Today, I'll make some *masala* tea," said Muni.

Ankita offered, "Let me join you."

REFLECT

1. Lifelong learning is the key to successful and sustainable leadership.
2. A successful leader harvests happiness alongside their team members.
3. Design thinking helps leaders be pragmatic, strategic and focused.
4. An imaginative leader harnesses the power of intuition alongside information, data and knowledge.
5. Leadership is a journey, not a destination.

YOUR LEADERSHIP QUESTIONS

1. What do you think are the basic elements of "sustainable" leadership?

2. How does design thinking help a leader enhance the shared vision in an organisation?

3. "Enlightened leadership is spiritual." How would you interpret and qualify this statement?

4. "You cannot connect the dots looking backward." How would you justify this statement?

5. "Designers do more than make things pretty." What are the other key features of design thinking?

18
Celebrating the Leader in the Self

*The vegetable life does not content itself
with casting from the flower or the tree a single seed,
but it fills the air and earth with a prodigality of seeds, that,
if thousands perish, thousands may plant themselves, that
hundreds may come up, that tens may live to maturity; that, at
least one may replace the parent.*
—Ralph Waldo Emerson
American essayist

After breakfast, as we settled down for a chat, I noticed three teenage girls approaching the house.

"Sir, may we come in?" they asked from the gate.

They were all dressed as though it were a festival day.

Seeing them from a distance, Ankita remarked, "Oh, today is *Aadi Perukku*."

Ankita welcomed the girls inside. "You are Kavita, right?" she asked one of the girls.

"Yes, I'm Shalini's daughter. My mother has sent this for you." She handed a small tiffin box to Viswa.

"What's inside?" Viswa asked, curious.

"Uncle, it's coconut rice," Kavita said.

"This is from my house. I'm Amuda's daughter. She has sent Sweet Pongal, made of rice, *moong dal*, ghee, jaggery, cardamoms and nuts," said the second girl.

"And I'm Swetha, Roopa's daughter. This is tamarind rice," said the third girl.

"Oh my God I don't think we need to prepare anything for lunch today," Viswa said.

"I think this evening is going to be special by the riverside," Ankita said.

I was excited. "I've never seen such goodwill among people in the village."

"That's how it is, Bala," Viswa said. "All their mothers attend my classes on financial literacy. They know I'm a loner. However, they take joy in sending special dishes on festivals to others. I'm just one of them."

"Festivals are occasions to foster camaraderie among people. This evening, all the villagers will assemble by the riverside to celebrate," he added.

"Sir, I've heard of Pongal as the harvest festival, but what about *Aadi Perukku*?" That was Amar.

"This festival is usually celebrated on the 18th day of the month called *Aadi Masam*, which usually falls between mid-July and mid-August. It marks the start of the monsoon and the arrival of fresh water flowing into the river. Farmers and their families gather to worship the local goddess and celebrate the arrival of rain. Let's go there this evening and see for ourselves," said Viswa.

He continued, "Ankita, I have some Mysore Pak sweets in the refrigerator. Could you please offer them to these girls?"

Muni said, "There's a dance programme near the river this evening. Our dance teacher has prepared about 10 girls for the performance."

"Uncle, there will also be a *kaviarangam* (poets' meet). My father is participating," one of the girls added.

We could already imagine the vibrancy of the evening. In the next few minutes, more dishes arrived for Viswa from other people. It was clear to see the geniality among the villagers, regardless of their religion, caste or community.

"All farmers join in this programme to express their gratitude to the river," Viswa continued.

I said, "Many conferences are held worldwide to raise awareness about the ecosystem and our relationship with Mother Nature. I believe these are the occasions that validate our connection and cohabitation with nature."

Viswa nodded and said, "What's crucial is that involving the village children and students of various age groups helps sow the seeds of these principles in their formative years. Such festivals echo several directives from the United Nations and its agencies."

When we reached the riverside around 6 p.m., it was bustling with activity. Many people, mostly mothers and daughters, were floating *diya*s (clay cups with lit cotton wicks) in the river, creating a beautiful sight. Muni stood by the water, hands folded, offering his prayers to the *diya*s.

"By the way, Viswa, who organises this event?" I asked, curious.

Viswa laughed heartily. "No one. People come with their families on their own. The sense of goodwill and comfort they

share with each other is amazing. This is a perfect example of how culture can bring us together."

"I see a stream of women carrying wooden baskets on their heads. What's inside the baskets?" I asked, intrigued.

"I think Muni can explain it better," Viswa replied.

Muni said, "The women in the village collect nine different grains, place them in mud pots, sprinkle water on them, and finally offer them to the river goddess as a mark of gratitude and respect. As they walk towards the river, they sing folk songs in praise of their goddess."

"Collective leadership, collaborative leadership and participative leadership—we can learn many lessons from these people and these events," I observed.

"You can't tell who's rich or poor here, or who's educated or uneducated. You can't even tell who belongs to which religion," Muni said.

"The spirit of leadership is visible in everyone, including the children," Ankita added.

"The interesting thing about farmers across the country is their ability to live close to nature, to be sensitive to it, to appreciate it, and to feel a deep connection with it. A number of harvest festivals, each with its own name, are celebrated everywhere with music, dance, and games, showcasing the camaraderie that farmers and their families share," Viswa said.

"Leadership is not just about leading others. It's also about leading oneself—every minute, every day and in every aspect," he added.

"Sir, there is a 'Rekla' race around that corner," Amar said. "It's a century-old bullock cart race. Once, Muni was the champion in that race."

"Wow! This is a vibrant village with countless leaders!" I exclaimed.

A middle-aged lady walked in and greeted Viswa. He greeted her back and introduced her to me. "This is Dr Bhargavi, the medical officer of this village. Since her arrival, health awareness programmes have been introduced, and people are following healthier practices."

"But the credit goes to Viswa sir," Dr Bhargavi said. "The positivity he's brought to this village is truly impressive. People follow him because they love and respect him. I was reading a book on leadership by John C. Maxwell, and he quotes Leroy Eims, a writer who also worked in various ministries in the United States: 'A leader is one who sees more than others see, who sees farther than others see, and who sees before others do'."

Viswa seemed uncomfortable with the praise. "Thank you, Doctor, but we are all different parts of the same universe, each contributing to the other."

"We must conduct leadership camps in the village for the corporates," Shyam observed.

Everyone agreed. Ankita added, "And Muni will be the coordinator of the programme!"

REFLECT

1. It is important to celebrate the power of leadership in every individual.
2. Cultural camaraderie triggers and catalyses social cohesion.
3. Life offers much more than money, wealth and professional growth.
4. Successful leaders focus on encouraging others to become leaders.
5. Leadership is the process of seeing the self in everyone they interact with and seeing those people in the self.

YOUR LEADERSHIP QUESTIONS

1. How does cultivating sensitivity towards the ecosystem help advance the goals defined by the United Nations?

2. Cultural cohesion is a powerful tool for community leadership. Discuss this statement.

3. "A leader is one who sees more than others see." Identify three perspectives or qualities of leaders that support this idea.

4. Why should a leader avoid being merely a drop in the ocean? What does this metaphor signify?

5. "Leadership is a process of seeing the self in everyone." What strategies can leaders use to put this idea into practice?

19
Gratitude: The Way of Life

*If your actions inspire others to dream more, learn more,
do more and become more, you are a leader.*
—John Quincy Adams
Former President (sixth) of the United States

We were all sitting together, quietly sipping strong coffee. "So, have you decided to go back today?" Viswa asked me, breaking the silence.

I smiled and said, "Yes, there's a lot of work waiting for me back home."

Ankita looked at us and smiled.

I responded, "Ankita, I sense there's a hidden meaning in your smile."

She replied, "Yes, sir. I've been observing both of you for the last three days. It seemed to me like you were playing a game of shuttle badminton, standing on opposite sides of the court. The shuttle was your thoughts, ideas, words and styles. Perfect shots, perfect moves, perfect placements…." Everyone laughed.

"As such, a conversation between two people is often compared to a dance," I commented.

Viswa nodded and added, "Bala's right. Social psychologist James W. Pennebaker wrote an interesting book called *The Secret Life of Pronouns: What Our Words Say About Us*. In it, he says, 'Conversations are like dances. Two people effortlessly move in step with one another, usually anticipating the other person's next move. If one of the dancers moves in an unexpected direction, the other typically adapts and builds on the new approach. As with dancing, it is often difficult to tell who is leading and who is following in that the two people are constantly affecting each other. And once the dance begins, it is almost impossible for one person to singly dictate the couple's movement.'"

"And what about leadership? What kind of dance is that?" Shyam asked, eager to jump in.

"That's exactly what we're doing," Viswa replied. "It's a group dance, harmoniously choreographed by the leader, with their mastery reflected in the movements of each dancer on the stage."

Shifting the conversation, Amar said, "I hope, Bala sir, you had a good time here in the village."

"Certainly," I replied. "My understanding of farming and farmland has completely changed. I never realised I was walking into a university where I'd learn so many lessons. Every patch of soil felt like a book in a library, and every farming operation was a discipline, with an enriched faculty, brimming with the passion to teach. I didn't just smell the soil; I could sense the aroma of change and transformation, from sowing seeds to harvesting."

As two people entered the house, Viswa stepped out to welcome them. All of us followed him and sat on the veranda.

Amar, Shyam and Ankita touched their feet as a mark of respect.

"God bless you, children," said one of the men, who was wearing a *dhoti* with a towel flung over his shoulder. He handed flowers and fruits to Viswa.

"Thank you so much," Viswa replied, then introduced him to me. "This is the temple priest. We have some beautiful temples in the village devoted to deities like Ganesh, Krishna and Shiva."

He then introduced the other person, who appeared to be in his 50s. "This is the local school headmaster Mr Muthiah, a great asset to the village. Many of his students have scaled high positions in life."

"I didn't do anything. They grew up on their own," Muthiah said humbly.

"That's his humility speaking—true words of a teacher," Viswa remarked. "He might not have taught them physics, chemistry or economics, but he taught them how to see life and how to handle it. He held their hands in their formative years so that they would grow in the right direction. He should be proud to have shaped great leaders."

"Teachers do that," Ankita said.

"Unfortunately, we fail to understand the values and contributions of the teaching community in developing leaders for the country," she added.

Muthiah smiled and responded, "Just as we often overlook the role of farmers in shaping the leadership of the country. They quietly play their part in forming leaders."

Mr Sankaraiah, the priest, nodded in agreement. He said, "Leadership is all about spirituality—moving from 'I' to 'We'; it's a continuous and sustainable process of growth together. It's how we look at ourselves and see ourselves in each other and in everything around us. We see our own reflection in others and others in us. That's the essence of what all the scriptures say. That's the joy of this cosmic exuberance."

We sat there in stunned silence. His words carried a deep meaning.

"A very powerful explanation," commented Amar.

"It's all about the process of re-engineering the self and continuously evolving," Muthiah added.

"Re-engineering the self?" Shyam questioned, intrigued.

Muthiah responded, "Shyam, you've studied photosynthesis in school, haven't you? What is it? The chlorophyll in the plant absorbs sunlight and uses water and carbon dioxide to make its own food, which is glucose. In a similar way, we should be able to receive light from wherever it is available and use our intellect and every other resource to produce our own energy, wisdom and growth potential. We need to do what plants do."

"*Wow! What a teacher he must be to put complex meanings into simple words!*" I wondered.

"Plants use carbon dioxide. But we can use our intellect to reduce carbon in our atmosphere to keep the planet alive. That is what the United Nations wants," Amar intervened.

"An effective leader is one who can use even waste materials and convert them into potential resources, Muthiah continued. "What do we do with garbage? Don't we convert it into biological resources? Similarly, in every system and every community, there are people who may seem unimportant or overlooked. They may not be as useless as we think. Perhaps their potential just hasn't been understood or used correctly. That's what a leader does—and that's what a teacher does in a classroom—brings out the potential."

Muni came in with two cups of coffee for the visitors.

"What about all of you?" Muthiah asked.

"We just had coffee," Ankita replied.

Muthiah turned to me and said, "Sir, we understand you've been in our village for the past three days as Viswa's guest."

Mr Sankariah added, "No, as the guest of the village." Everyone smiled.

He added, "Today, around 4 p.m., you're invited to a small *gram sabha*, a get-together organised by the villagers to honour you and get your feedback about our village. We came to request you to join us."

When Muni looked at me with a smile, I could tell he had a hand in this invitation. I was deeply touched.

"Of course, sir," I said. "It would be an honour to be with all of you."

Later in the afternoon, as I was packing my luggage, Ankita walked in with a package. "Sir, this is for you," she said as she handed it to me.

"What is this?" I asked, curious.

"This is *papad* made by the village community. Inside the packet, there's a herbal powder that can be taken with boiled water or honey to treat a cold or cough. There's also some export-quality pure turmeric powder made in our village, and in the bottle is pure honey, another product of our cottage industry."

I thanked her profusely. My perception of the village and farming had undergone a massive shift. Small-scale and cottage industries, homemade products and other food items, had been served to several lakh people without any branding, marketing or profit-making.

"Leadership is not about leading big brands or earning a high corporate salary. It is also about leading at all levels and contributing to the self and community for growth and well-being," I recalled Viswa's words.

By 4 p.m., as we entered the community hall of the village, I noticed nearly a hundred people had gathered—both men and women, sitting quietly. Viswa and I were offered seats on the stage alongside the *gram sabha* chairman and the *panchayat* president. A lady sitting with us was the head of the village women entrepreneurs' council.

Sankariah began with the traditional *Athithi devo bhava*— a greeting meaning 'the guest is like a god', as prescribed in our scriptures.

He then spoke, "Hospitality is a part of our culture. We've had a guest with us in the village for the last three days—thanks to our Viswa sir for bringing him here. Viswa has been our friend, philosopher and guide. He has given us a vision and energy. He has taught us about the meaning of life and how to seek happiness amidst difficulties through collective wisdom. I'm sure our guest has admired our village, but we have nothing much to teach the world. However, we have learnt that every village has a lot to offer the world: food for the stomach, for thoughts and for happiness. I hope you have seen some elements of that in our village."

The *panchayat* president presented me with a shawl, while the head of the women's entrepreneurs' council gifted me a small bottle of jasmine essence. When it was my turn to speak, I stood up and said, "I'm not a bad speaker, but my words are failing

me now. When love speaks, words celebrate silence. When actions speak, words stand listening. When wellness and peace speak, words become students. Perhaps that's why my words are struggling to perform in an atmosphere of love, action, positivity, beauty and peace. I cannot express my admiration for Viswa, my friend. He has become a *Karma Yogi* through his selfless actions to benefit those around him, without expecting any reward. I can only offer my appreciation for the young leaders he is nurturing and mentoring—Amar, Ankita and Shyam. The world needs people like them. And Muni, a leader for all seasons and occasions, has shown me that leading is not exclusively about academics. It is so much more."

As I walked among the people there, in my mind I saw seeds filled with love, the leaves of trees dancing to the breath of the people, the cattle responding to their calls like family members. Interestingly, I saw no fear or anxiety in anyone's expression. I breathed in the air, which carried the scent of unconditional love and selfless service. What more could a community need? The village is an open university, and every farmer here is a teacher. I salute all of them.

I could only say, "Thank you all. I hope other villages learn from you, and other leaders follow your exemplary leadership. I've thoroughly enjoyed my stay and will certainly come back soon to learn more."

Despite many requests, Viswa declined to speak, apologising with folded hands. He knew his actions spoke louder than his words.

Muthiah proposed the vote of thanks. He said, "One may wonder why we organised this meeting. I recall the famous words of the great saint and Tamil poet, Thiruvalluvar, who says in one of his poems, 'Like the Anicham flower which shrinks on smelling, the guests also turn a sour face when we turn away our faces from them'. This is why many of our own people turn away from villages—they feel unacknowledged, and their productive work is not appreciated. Many of our young people prefer towns and cities, without realising how villages can contribute to their growth, well-being, health and peace. We are fortunate to have youth here like Amar, Shyam and Ankita. I'm sure our guest will take green memories of this village, along with his belief that our country still thrives in the vibrancy of villages and the leadership of farmers. While we certainly acknowledge that the whole world is changing, going digital and replacing old ways with new, the leadership lessons people get through farming are just as vital as those gained in colleges and universities. We continue to be the laboratory for leadership practices."

Everyone clapped, wholeheartedly agreeing with his views.

While I was gathering my luggage from the room, I called out to Viswa, "I've written a cheque in favour of your village welfare society. It's a donation from the trust I established in my father's name. Please use this for the betterment of the schools and hospitals here."

Viswa looked at the cheque and exclaimed, "This is a huge amount—Rs 10 lakhs!"

I smiled, though a sense of sadness lingered as I prepared to say my goodbyes. I could see tears in the eyes of Ankita, Amar and Shyam, and I remembered the famous words of Samuel Taylor Coleridge: "To meet, to know, to love, and then to part, is the sad tale of many a heart". Muni was waiting outside with his bullock cart to take me to the railway station.

Ankita's eyes were moist as she asked, "Sir, when is your next visit?"

I smiled and replied, "As soon as I finish writing my book, *A Walk with Viswa*.

Despite the sadness of parting, all of them smiled.

> **REFLECT**
>
> 1. Meaningful conversations lead to understanding and empowering relationships.
>
> 2. A leader's performance profile is valued more than the personal brand they cultivate.
>
> 3. Recognition and respect for strong leadership naturally follow the leader's contributions to the community.
>
> 4. A leader is a source of positivity for their team members.
>
> 5. An effective leader transforms individual energy into collective synergy.

YOUR LEADERSHIP QUESTIONS

1. What are some important features of a good conversation?

2. "Every bit of the soil looked like a book in the library." What are the three lessons one could learn from the soil?

3. Leadership is a process of moving from "I" to "We". Suggest two strategies to accelerate this process.

4. Why should a leader carry positivity on their shoulders?

5. How and when does the transformation of individual energy into synergy occur? Provide an example.

Epilogue

Every beginning has an end and that end often sets the tone for a new beginning. My visit to the village was not a full stop, but merely a comma. It quietly sowed the seeds for many new experiences that I will soon write about. But perhaps the most significant impact of my stay in the village was the reaffirmation of my belief that India—Bharat—truly lives in her villages.

Every village in India breathes the spirit of Bharat, and with every breath comes the warmth of a human generation filled with aspirations—for love, peace, co-existence and a vision to see a world built on caring and sharing. These villages embody the ancient wisdom: "The whole world is one family."

Each visit revealed to me the immense, often unspoken, wisdom embedded in the human capital of these communities. It dispelled the myth that intellect alone is sufficient for a meaningful life. I came to realise that the language of the heart is far more powerful than the most eloquent words found in any dictionary.

Each time I returned to the village, I felt compelled to touch the soil through which Mother Earth whispered a culture and heritage of several centuries. I began to feel her boundless compassion and love for all her children, transcending borders and reaching beyond geographies.

I started looking at each tree, each leaf and each flower as the fruit of the labour of the wise souls over centuries who shaped this land into what it is today. Through them echoed the stories, poems, folklore, music, dance, biographies and the history of generations who had lived there for centuries.

I felt nursed and nurtured by the five elements through which Mother Earth spread her love and order to the people she had given the freedom to live on her soil for centuries.

I came to understand that every farmer in the villages of Bharat is an artist in their own rights, painting their imagination on the canvas of their dreams. There was no reason for me to consider them inferior to any other talented person on the planet who claimed advanced insight into the universe of knowledge.

Viswa became immortal—not just as a person, but as an idea. Oscar Wilde, the noted literary genius, once said: "The value of an idea has nothing whatever to do with the sincerity of the man who expresses it." In Viswa's case, however, his sincerity was unquestionable. His teachings on why and how we should live than merely survive helped distance me from my fears and anxieties.

Along with others, I accepted Muni as a *guru* who taught me about life more through his silent presence than his words. The three young learners, who were trying to discover themselves through their actions, showed me that opportunities for learning exist in every situation, if only we have the will and openness to receive them.

The village taught me that there is life in the breeze, dopamine in each moving leaf and joy in every fruit that blossom.

I saw friendliness in every animal and hope in the eyes of every human being.

The farmers look up from their soil and wonder what lies beyond the sky. On the other hand, NASA astronaut Sunita Williams looked down at Mother Earth from space and said, "My space expedition has changed my perspective towards people. Looking down at the Earth, we could not see borders or people with different nationalities. It was then that realization dawned on us that all of us are a group of human beings and citizens of the universe."

Human consciousness manifests in different ways, in different contexts. Yet at its core, we are one. There is deep meaning in the ancient wisdom of "*Vasudhaiva Kutumbakam*" (the whole world is one family).

So, what more should we look forward to?

Let us stay happy and healthy, and celebrate life until we meet again!

About the Author

G. BALASUBRAMANIAN, a native of Tirunelveli, Tamil Nadu, began his professional journey as an educator and rose through the ranks to become the Director (Academics) of the Central Board of Secondary Education, where he served for over two decades. His interests span across literature, education, philosophy and psychology.

A seasoned professional trainer, he has conducted over 5,000 workshops for teachers, educational administrators, institutional leaders and corporate professionals.

Mr Balasubramanian is the author of several influential works for educators and administrators, including *Mindscaping Education, Case Studies in Classrooms, Quality Spectrum: A School's Bandwith* and *Safety in Schools*, among others.

His literary works include poetry collections *A Happy Journey* and *Apologies to an Existence*.

He has been honoured with numerous accolades, including the Seva Ratna Award from the Centenarian Trust, Chennai, and the Lifetime Excellence Award from VIMHANS, Delhi, in recognition of his outstanding contributions to adolescent education.